ST(P) Te▮▮▮
A Series▮▮▮

−10.99

ENGINEERING INSTRUMENTATION AND CONTROL

Second Edition

D C Ramsay, BA(Hons), MIMechE

Stanley Thornes (Publishers) Ltd

First published in 1981 by
Stanley Thornes (Publishers) Ltd
Old Station Drive
Leckhampton
CHELTENHAM
Glos. GL53 0DN
England

2nd Edition 1984

Reprinted 1986, 1991, 1993

British Library Cataloguing in Publication Data

Ramsay, D.C.
  Engineering instrumentation and control.
  2nd ed.
  1. Automatic control   2. Engineering instruments
  3. Measuring instruments
  I. Title
  629.8   TJ213

  ISBN 0-85950-225-2

Typeset by Tech-Set, Gateshead, Tyne & Wear.
Printed and bound in Great Britain at The Bath Press, Avon.

Contents

Preface to the Second Edition

This book was originally written to cover the Technician Education Council's course unit *U77/422: Engineering Instrumentation and Control IV*. The BTEC have since replaced that unit with a later version, *U83/122*, and I have therefore, in this second edition, made such additions and deletions as are necessary to ensure that it covers the new BTEC unit without being unduly lengthy.

The main difference is that more attention is now paid to the electrical and electronic aspects of instrumentation and control, and less to the 'hardware' aspects of it — the gear systems, the lever systems, and mechanical equipment in general.

Because of this, and because most mechanical engineering students seem to have difficulty in comprehending the electronic side of instrumentation, I have completely rewritten Chapter 6, the chapter on measurement projects. This chapter is now devoted to one project only, the building and testing of a pair of simple amplifiers and the determination of their frequency response, singly and in series. Enough detailed information is given for any student to be able successfully to complete such a project, and the experience, confidence and sense of achievement which result should prove extremely valuable.

Gloucester, 1984   DCR

Acknowledgements

The author and publishers wish to thank the following for permission to copy or reproduce illustrations:

Central Electricity Generating Board, Leeds: Fig. 1.1

Gould Advance Ltd, Hainault, Essex: Fig. 4.5

R S Components Ltd, London: Figs. 3.7 and 3.8

Southern Measuring Instruments Ltd, Sandown, I.O.W.: Figs. 4.2 and 4.3

They also wish to thank The Institute of Quality Assurance, London, for permission to use those examination questions that have been appended '(I.Q.A.)' and Tek Quipment International Ltd, Nottingham for their help in providing the cover photograph.

Symbols and SI Units

SYMBOLS FOR PHYSICAL QUANTITIES

The following is a list of symbols used in this book, which are generally in accordance with *British Standard* 1991:

A area
C capacitance
d relative density
e the exponential, the base of natural logarithms ($= 2.718\ldots$)
F force
f frequency
g acceleration of a freely falling body at Earth's surface ($9.81\,\text{m/s}^2$)
h height, head of liquid
k gain, gauge factor
l length
p pressure
$\dot{Q}$ volume flow rate
q charge
R resistance
s displacement, distance travelled
T absolute temperature, period
t time
V voltage (constant)
v voltage (varying)
X maximum displacement during a half-cycle of vibration
x displacement
α (alpha)   angle of inclination
δ (delta)   the natural log of the ratio of successive half-cycle amplitudes of a vibration
ϵ (epsilon)   tensile or compressive strain
ζ (zeta)   damping ratio
θ (theta)   angle turned through
λ (lambda)   spring rate
ν (nu)   Poisson's ratio
π (pi)   ratio of circumference to diameter of a circle ($3.14\ldots$)
ρ (rho)   density
τ (tau)   time constant
ϕ (phi)   phase lag
ω (omega)   angular velocity

SYMBOLS FOR SI UNITS

SI stands for the Système International d'Unités, or International System of Units, which has been adopted by British industry for engineering calculations.

The following is a list of symbols used in this book which are also in accordance with British Standard 1991:

A   ampere
bar   bar   ($= 10^5 \, N/m^2$)
C   coulomb
°C   degree Celsius
F   farad
Hz   hertz (cycle per second)
K   kelvin (absolute °C)
kg   kilogram(me)
m   metre
min   minute
N   newton
Pa   pascal   ($= N/m^2$)
rev   revolution
s   second
V   volt
Ω   ohm

To express multiplies and subdivisions of SI units concisely, the following prefixes are used with them:

Multiplication Factor		Prefix	Symbol
1 000 000 000	$= 10^9$	giga	G
1 000 000	$= 10^6$	mega	M
1 000	$= 10^3$	kilo	k
0.01	$= 10^{-2}$	centi	c
0.001	$= 10^{-3}$	milli	m
0.000 001	$= 10^{-6}$	micro	μ
0.000 000 001	$= 10^{-9}$	nano	n
0.000 000 000 001	$= 10^{-12}$	pico	p

SYMBOLS FOR ELECTRICAL COMPONENTS

These are generally in accordance with BS 3939: *Graphical Symbols for Electrical and Electronics Diagrams*. Resistors are indicated by the objective symbol ⎯▭⎯ except where there is a possibility of its being confused with a block diagram rectangle. In such cases the alternative symbol ⎯/\\/\\/\\⎯ is used.

ABBREVIATIONS

The following abbreviations and contractions are used in this book:

a.c.	alternating current
AM	amplitude modulation
d.c.	direct current
deg	degrees of temperature
d.i.l.	dual-in-line
FM	frequency modulation
f.s.d.	full-scale deflection
i.c. op. amp.	integrated circuit operational amplifier
ln	natural logarithm, log to the base e
LVDT	linear variable differential transformer
mm Hg	millimetres of mercury
mm H_2O	millimetres of water
m.p.h.	miles per hour
PVC	polyvinyl chloride
VHF	very high frequency
UV	ultraviolet
δ	a small change of

PART ONE

Instrumentation

Chapter 1

Measurement Systems

ANALYSIS OF A SYSTEM

Engineering depends upon measurement. To manufacture engineering components, or to control continuous processes in, say, a power station, or to test vehicles, engines or structures, a great deal of precise information must be obtained — by measurement.

An electricity generating station, for instance, is controlled by engineers who operate control consoles like that shown in Fig. 1.1, containing literally hundreds of indicating dials. These form a continual display of measurements of such things as the flow rates of fuel and air to the burners, feed-water to the boilers and cooling water through the condensers, together with pressure and temperature readings at many vital points throughout the plant. Each dial displays the output of a separate measurement *system*.

Fig. 1.1 Control console of CEGB 660 MW generating set. *Source:* CEGB, Leeds Training Centre

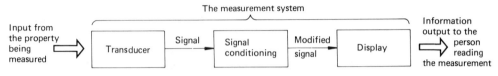

Fig. 1.2 Block diagram

Each measurement system in such an array can be represented by the block diagram shown in Fig. 1.2.

A *transducer* is a device which converts a property difficult to measure into another property more easily measured. For example, suppose we want to measure the temperature of water. The temperature of a substance depends on the intensity of vibration of its atoms and molecules. We cannot hope to measure the vibrations of particles so small that they can never be seen, but if we use an ordinary thermometer, we are making use of a transducer, in the form of a bulb full of mercury, which converts the vibrations into an expansion or contraction of volume. This is much easier to deal with.

The output of a transducer is a *signal*. This contains the essential information on the measurement we are trying to make. Often a signal is in the form of an electrical voltage or current, or an air pressure or liquid pressure, but in the case of our ordinary thermometer, the signal is a change of volume.

The change of volume is quite precise, but far too small to be of any use if the mercury remains in the shape of the bulb. So *signal conditioning* is required, to change the nature of the signal in some way. In the case of our thermometer, the change of volume of the mercury is channelled along the capillary tube formed in the glass stem, so that the change of volume becomes a change of length in the thread of mercury which can be seen through the glass.

The signal in its final form must be *displayed*, so that a reading can be taken, and this is done directly, in the case of the thermometer, by comparing the end of the thread of mercury with the degree graduations on the stem.

I have chosen a simple thermometer to illustrate the essentia features of a measurement system, but *any* measurement system consists of only these same three essentials: *transducer, signal conditioning* and *display*. The difference, in the case of the more complex systems such as those shown in Fig. 1.1, is that the signal conditioning block in Fig. 1.2 can be split up into a number of component blocks. For example, a temperature measurement system designed to record its measurements automatically in the form of a graph might consist of: (a) a thermocouple, (b) an amplifier, to make the thermocouple output powerful enough to operate a chart recorder, (c) a low-pass filter circuit, to filter out any mains frequency ('mains hum') that might be picked up by the connecting cables, and (d) the chart recorder, to plot the graph of temperature against time. The block diagram of such a system would then be as in Fig. 1.3.

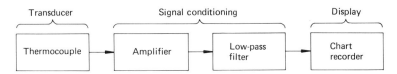

Fig. 1.3 Block diagram for a temperature measurement system

For further examples of how measurement system block diagrams may be drawn up, and how they may be used to calculate values of signal gain required, I am going to look at two common measuring devices: the pressure gauge and the moving-coil meter. These often occur as display devices in more complicated measurement systems and thus can be thought of as fundamental 'building blocks' of measurement systems in general.

Other examples of the use of block diagrams to represent measurement systems or their components will be found on pp. 28, 30, 43, 117, 118 and 121.

THE BOURDON TUBE PRESSURE GAUGE

Fig. 1.4 A Bourdon tube pressure gauge

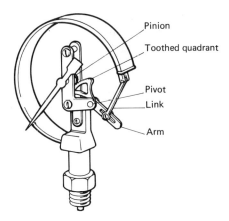

The standard type of pressure gauge, shown in Fig. 1.4, consists of a Bourdon tube driving a pointer, through gearing. The Bourdon tube is a tube of oval cross-section, bent into an arc of a circle and closed at one end. When pressure is admitted to the other end of the tube, the oval section is blown out into a more circular cross-section, and this causes the tube itself to tend to straighten out so that it becomes an arc of greater radius (Fig. 1.5). So the Bourdon tube is the transducer, converting pressure to displacement.

Fig. 1.5 The Bourdon
tube

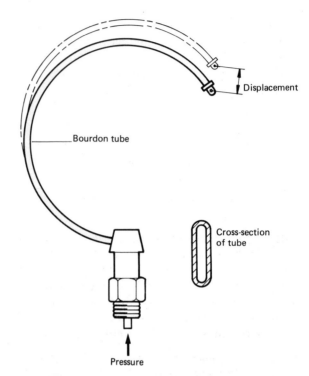

However, the movement of the tip of the tube (the signal) is almost imperceptible, so it has to be amplified (signal conditioning). The amplifier in this case is mechanical, a toothed quadrant and pinion, but it amplifies angular, not linear displacement. We therefore need to convert the signal from a linear to an angular displacement, and this is done by linking the tip of the Bourdon tube to an arm on the opposite side of the toothed quadrant (see Fig. 1.4).

Finally, having obtained the amplified signal as a rotation of the pinion, we have to display the result, and this we can easily do by fitting a pointer to the end of the pinion spindle so that it points to the pressure reading on a circular scale. The block diagram of the pressure gauge is therefore as shown in Fig. 1.6.

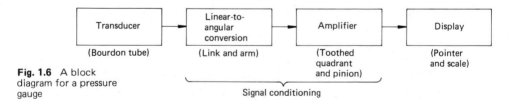

Fig. 1.6 A block
diagram for a pressure
gauge

If we insert into each block the relationship between its output and input, we can use the block diagram to design a measurement system, because the overall relationship between input and output is obtained by multiplying together the individual relationships which are in series.

EXAMPLE   A pressure gauge is to be designed to have a pointer rotation of 300° when the pressure is varied from 0 to 10 bar. The Bourdon tube deflects 2.5 mm at the tip connection, for a pressure increase of 10 bar. It is linked to the arm at a radius of 15 mm. Determine a suitable gear ratio for the quadrant and pinion.

SOLUTION   The block diagram is drawn with the values of $\frac{\text{output}}{\text{input}}$ (that is, the signal gain) written into each box, the unknown gear ratio being denoted by a gain of k. The gains of the other components are

Bourdon tube

$$\frac{\text{Output}}{\text{Input}} = \frac{2.5}{10} = 0.25 \frac{\text{mm}}{\text{bar}}$$

Link and arm

The movement is assumed to be small enough for the link to be taken as tangential to the connecting pin's arc of travel, throughout its range. Then for a link movement of s mm, the angle turned through by the arm is $s/15$ radians, as shown in Fig. 1.7.

Fig. 1.7 See example

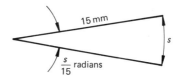

$s/15$ radians is the same as $\frac{s}{15} \times \frac{180°}{\pi} = 3.82s$ degrees.

Then the gain is

$$\frac{\text{Output}}{\text{Input}} = \frac{3.82 s}{s} = 3.82 \frac{\text{deg}}{\text{mm}}$$

Pointer and scale

This is purely a display device, and it can be considered as having a gain of 1 — that is, the output is $1 \times$ the input.

The block diagram with numerical values inserted is therefore as shown in Fig. 1.8.

Fig. 1.8 Block diagram with values of gain inserted

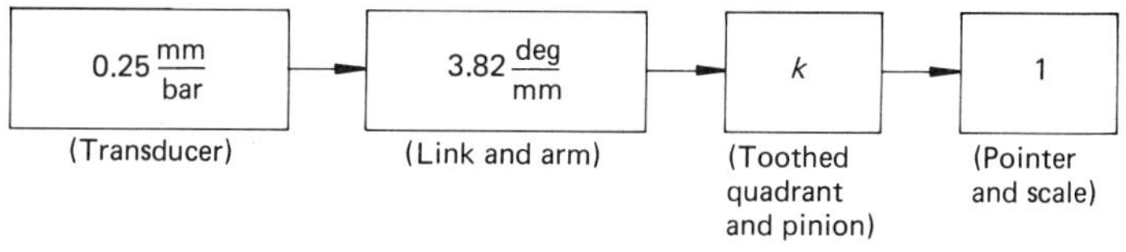

$0.25 \frac{\text{mm}}{\text{bar}}$	$3.82 \frac{\text{deg}}{\text{mm}}$	k	1
(Transducer)	(Link and arm)	(Toothed quadrant and pinion)	(Pointer and scale)

The overall output/input relationship is called the *sensitivity* or *scale factor* of the instrument. In this case the sensitivity is to be

$$\frac{300°}{10 \text{ bar}} = 30\frac{\text{deg}}{\text{bar}}$$

Then multiplying the signal gains together gives

$$0.25 \times 3.82 \times k \times 1\ \frac{\text{mm}}{\text{bar}} \times \frac{\text{deg}}{\text{mm}} = 30\frac{\text{deg}}{\text{bar}}$$

$$\therefore \qquad k = \frac{30}{0.25 \times 3.82} = 31.4$$

Thus a standard gear ratio of 30 to 1 would be suitable, though this would give a slightly smaller pointer rotation than the 300° specified. To put this right, we could slightly modify the radius at which the link is pinned to the arm (that is what the slot is there for).

When we obtained the gain of the link and arm, we *divided* by the radius (15 mm), so now we should reverse that operation (i.e. *multiply* by 15) and divide instead by a new radius, x mm.

Then to neutralise the effect of taking gear ratio k as 30 instead of 31.4 we can say

$$\frac{30}{31.4} \times \frac{15}{x} = 1 \quad \text{(because we want the effects of the two changes to cancel out)}$$

$$\therefore \qquad x = \frac{30 \times 15}{31.4} = 14.3 \text{ mm}$$

$$Check: \quad 0.25 \times \frac{180}{14.3 \times \pi} \times 30 \times 1 = 30\frac{\text{deg}}{\text{bar}}$$

THE MOVING-COIL METER

This is the other commonly used display device, which can also be treated as a measurement system in its own right. Its construction is shown in Fig. 1.9.

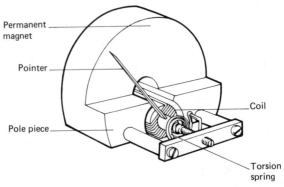

Fig. 1.9 Moving-coil meter

Permanent magnet

Pointer

Pole piece

Coil

Torsion spring

Fig. 1.10 The coil
former

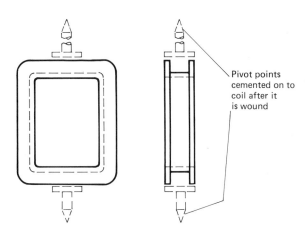

Pivot points
cemented on to
coil after it
is wound

The transducer is a coil of very fine enamelled wire wound on to a
rectangular aluminium coil-former (Fig. 1.10) and suspended so
that it is free to turn through about 90° in the magnetic field
between the poles of the permanent magnet. The gap between the
poles is circular, and a cylindrical core of soft iron is rigidly suspended, concentrically in the gap, to pull the magnetic field in, so
that it is virtually radial in relation to the centre of rotation of the
coil (see Fig. 1.11). The coil can then rotate in the gap between
the magnet poles and the iron core.

Fig. 1.11 The magnetic
field

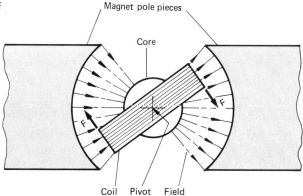

Magnet pole pieces

Core

Coil   Pivot   Field

The magnetic force which acts on an electrical conductor in a
magnetic field is proportional to the current flowing through the
conductor, and perpendicular to both the field and the current. So
the total force, F, acting on one side of the coil in Fig. 1.11 is
proportional to the current. On the other side of the coil the
direction of the field is unchanged but the current is here flowing
in the opposite direction, so an equal and opposite force F is
produced. The two forces form a *couple* giving a turning moment
proportional to current.

We now need a means of converting torque into angular displacement, and this is done by some kind of torsion spring. In the usual commercial meter, the spring is a spiral, similar to the hairspring of a (mechanical) watch or clock. In sensitive current-detecting meters (galvanometers) the ends of the coil are soldered to a phosphor bronze strip at top and bottom of the coil, which functions as (a) the top and bottom suspension points of the coil; (b) the torsion spring, and (c) a frictionless means of making the electrical connections to the coil (Fig. 1.12).

Fig. 1.12 Taut band suspension

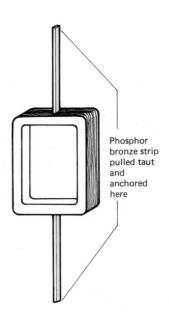

Phosphor bronze strip pulled taut and anchored here

The rotation of the coil against the resisting torque of the spring now needs to be *displayed*. In a simple commercial meter this is done by providing a radial pointer, carried by the coil, which indicates a value on a scale with radial graduations. Where the value is to be recorded rather than displayed, the pointer may be replaced by an arm carrying some kind of stylus or pen, to draw a graph of current against time on a paper strip unrolling at constant speed. This type of instrument is called a *chart recorder*. A pen-carrying arm has, however, a comparatively large moment of inertia, and this limits to about 10 Hz the highest frequency which it can follow with reasonable accuracy. To follow higher frequencies than this, the pointer or arm is replaced by a tiny concave mirror which reflects a ray of light (usually ultraviolet) on to a paper strip coated with a chemical which is sensitive to the reflected light, so that the trace develops gradually a few minutes after it has been exposed. This instrument is known as an ultraviolet recorder (described in more detail in Chapter 4).

Thus the block diagram of a moving-coil meter is as shown in Fig. 1.13.

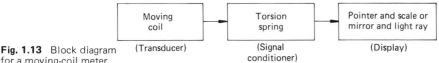

Fig. 1.13 Block diagram for a moving-coil meter

(Transducer)   (Signal conditioner)   (Display)

EXAMPLE   The coil of a moving-coil meter develops a torque of $0.002\,\text{N m}$ when a current of $500\,\mu\text{A}$ is passed through it. Determine a suitable stiffness for the torsion spring if the meter is to read $100\,\mu\text{A}$ at the full scale deflection of $90°$.

SOLUTION   The signal gain of the transducer is

$$\frac{\text{Output}}{\text{Input}} = \frac{0.002}{0.0005} = 4\,\frac{\text{N m}}{\text{A}}$$

The sensitivity of the instrument is to be

$$\frac{\text{Output}}{\text{Input}} = \frac{90}{0.0001} = 900\,000\,\frac{\text{degrees}}{\text{amp}}$$

Thus the block diagram with the numerical values written in is as shown in Fig. 1.14.

Fig. 1.14 See example

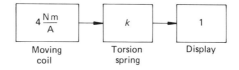

Moving coil   Torsion spring   Display

Therefore

$$4\,\frac{\text{N m}}{\text{A}} \times k \times 1 = 900\,000\,\frac{\text{deg}}{\text{A}}$$

$$\therefore \qquad k = \frac{900\,000}{4}\,\frac{\text{deg}}{\cancel{\text{A}}} \times \frac{\cancel{\text{A}}}{\text{N m}}$$

$$= 225\,000\,\frac{\text{deg}}{\text{N m}}$$

This is the spring constant, and from its units we can see that it is the reciprocal of the spring stiffness, the units of which should be $\frac{\text{load}}{\text{deflection}}$. The required spring stiffness is therefore

$$\frac{1}{225\,000} = 4.44 \times 10^{-6}\,\frac{\text{N m}}{\text{degree}}$$

VOLTMETERS AND AMMETERS

You will probably appreciate by now that the moving-coil meter is a sensitive, rather delicate current-measuring instrument. Because the wire of the coil is fine, it has a fairly high electrical resistance (300 Ω typical), and a very small current will cause the maximum reading to be reached (full-scale deflection). If we pass very much more than the full-scale deflection current through it, we are liable to melt the wire of the coil — the instrument would then be ruined. However, we can use the moving coil meter to measure large currents (i.e. as an ammeter) by bypassing most of the current through a shunt — that is, a very low resistance in parallel with the meter — and changing the numerical values of the scale to correspond.

Similarly, although the coil could be burnt out by quite a small voltage applied across its ends, we can use the meter to measure large voltages by putting a very high resistance in series with it, and suitably altering the scale, so that it is really responding to the small current which is being passed by the total resistance of coil plus series resistor.

These two variations are shown diagrammatically in Figs. 1.15 and 1.16.

Fig. 1.15 Construction of an ammeter

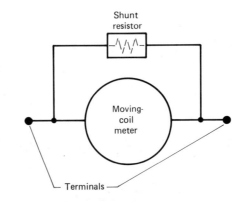

Fig. 1.16 Construction of a voltmeter

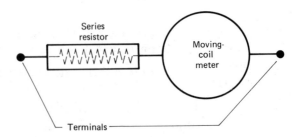

DEFINING THE QUALITIES OF A MEASUREMENT SYSTEM

Whenever we take a reading from an instrument, we can be sure of one thing — that we are not getting the right answer. Usually, the wrong value that it gives is so close to the true value that the difference (the *error*) is negligible, but error there will be, if we express the true value precisely enough. For instance, a steel rule may show us that the thickness of a metal block is 10 mm, whereas its true thickness may be 10.103 mm. Although there is nothing wrong with the rule, we have a measurement error of 0.103 mm. If an error of this magnitude is serious, we must have been using the wrong kind of measuring instrument — perhaps we should have used a micrometer.

Measurement system specifications therefore use various technical terms to describe or define the various kinds of error we may expect.

ACCURACY

Accuracy is usually specified in terms of error — thus the steel rule, as a measurement system, may be said to have an accuracy of ± 0.2 mm because, although it has probably been made to a far higher accuracy than that, the *resolution* of our unassisted eyesight is such that we cannot reliably estimate fractions of a millimetre finer than 0.2 mm. The accuracy of an electrical instrument such as a moving-coil meter is usually expressed as a percentage of the full-scale deflection (f.s.d.) value of the meter. Thus if the range of a voltmeter is from 0 to 10 volts, and the maximum possible error in its reading is 0.1 volt, then its accuracy is ± 1% f.s.d.

Quoting the error as a percentage of a full-scale deflection tends to obscure the fact that the possible error is a virtually constant number of scale divisions at any point on the scale — in this example a constant 0.1 volts. Thus if we take a reading at the lower end of the scale, say at 2.5 volts, the possible *percentage error on our reading* is $(0.1/2.5) \times 100\% = 4\%$, which is becoming unacceptably high. In such a case we should use a more sensitive instrument, or switch to a more sensitive range on the same instrument. Electrical test meters (multimeters) for example have such a facility. Ideally, the ranges should descend in ratios of about 3 to 1; e.g. 0–100, 0–30, 0–10, 0–3, 0–1, etc., so that we can always select a range such that our readings come within 0.25 to 0.9 of f.s.d. When using such an instrument, always switch in to the least sensitive range first (0–100 in the example above), and then downwards (if necessary) from that, to minimise the risk of damaging the instrument by overloading it.

In some instruments, the error may include a *zero error*, or *zero shift*, which means that when the input was zero, the pointer did not indicate zero but some value close to it. If it indicated a value of z units, then as well as all the other errors to which the instrument may be subject, an error of z units would be added to all readings. In a moving-coil meter such as that in Fig. 1.9, zero error can be eliminated — there is usually a small button with a screwdriver slot, set in the instrument casing, over the pivot of the coil, which zeros the pointer by adjusting the anchorage of the torsion spring. This should be adjusted, if necessary, before the instrument is used.

When the measurement is made as a comparison with a standard measurement, using some kind of comparator, an additional source of error is the allowable variation in the measured value of the standard; that is, the tolerance on the standard. An example of this is the use of gauge blocks (slip gauges) to determine a thickness measurement. To the maximum error in the comparator reading must be added the sum of the tolerances on the gauge blocks being used to make up the thickness. For this reason, standards such as gauge blocks are manufactured to very fine tolerances indeed, and a dimension should be made up using as few blocks as possible.

REPEATABILITY

Another source of error is due to random effects such as friction. When the same measurement is made over and over again, with the instrument allowed to return to zero each time, it is found that there is usually a slight variation in the readings. The *repeatability* or *reproducibility* of the instrument is the overall range of such variations when the instrument is thus tested over a short period of time under fixed conditions of use. It is often specified as a percentage of full-scale deflection.

STABILITY

Error may also be introduced by a change in the signal gain of the transducer or signal conditioning components of a measurement system. Electronic amplifiers, for instance, are rather liable to this fault, especially if they are operating at very high values of gain. The change of gain may be due to adverse environmental conditions; for instance, temperature, humidity, pressure, vibration, electric and magnetic fields, and power supply variations can all affect the gain. Or the change may be due to ageing of materials; this is called *secular change*, and even solid metal components are subject to it in the form of distortion caused by a gradual relaxation of stresses within the material. The ability of a system or a component to resist such changes is referred to as its *stability*.

The stability of a measuring instrument or system can be defined as its repeatability under specified conditions of use when the repeatability tests are separated by long periods of time.

We shall meet the word 'stability' again when we study automatic control systems. In that context, a control system has stability when oscillations, caused by disturbances, die out rather than increase in amplitude.

Constancy is a similar property to repeatability, but in this case the instrument or system is supplied with a continuous input of constant value, and the constancy is determined by the variation of its reading over a period of time while the conditions of the test are varied within specified limits.

CALIBRATION

When a measurement system is constructed, it has to be *calibrated* — that is, the relationship between the output of the system (the displayed value) and the input to the system (the quantity being measured) must be determined. Even instruments which are bought as already calibrated units, such as oscilloscopes or load cells, need regular recalibration because of the possibility that they may be affected by secular or environmental change. Only instruments whose calibration is unlikely to change because of the simplicity of their construction (for instance, glass thermometers) need no further calibration, and even simple instruments like these should have a recalibration check if there is any doubt or discrepancy.

Calibration is done by applying inputs to the system and reading off the corresponding output values. The output and input values are then plotted on a graph of 'displayed value' against 'input value'; this is the *calibration curve* for the system — which may well be a straight line if the response of the system is *linear*.

Ideally, the input should be increased in about ten equal stages covering the whole range of the instrument in one direction; it should then be returned to the starting value by similar stages in the reverse direction. The purpose of this procedure is to detect any *hysteresis* in the system. Hysteresis is the tendency of the output to remain at its original value when the direction of change of the input has been reversed. A typical cause is friction in mechanical parts. Fig. 1.17 shows a typical hysteresis loop which might be obtained in a calibration test. It will be seen that hysteresis introduces a further source of error; the output could be in error by an amount x depending on whether the input was increasing or decreasing. For this reason, input changes during calibration should be in one direction only, except when at the end of the range. Thus if we overshoot an increment during calibration we must read the output value wherever the input stops, even if it does result in unequal spacing between points on the calibration curve.

Fig. 1.17 A hysteresis loop from a calibration test

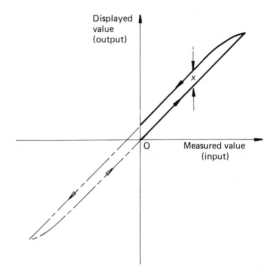

When an already calibrated system is to be given a check calibration, it may be preferable to plot error (displayed value minus true value) against displayed value, as in Fig. 1.18. This shows up zero error, nonlinearity, hysteresis and random errors on a greatly magnified scale.

Fig. 1.18 Typical graph of instrument error

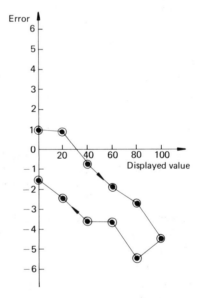

PRIMARY AND SECONDARY STANDARDS

When we calibrate a measurement system we must use precisely known input quantities. These may be obtained from a *primary standard* — that is, they may be produced using the means by which the quantity is defined; or we may use a *secondary standard* — that is, the input may be produced or measured by a system which has itself been calibrated using a primary standard, and which is of higher accuracy than the system we are calibrating.

For example, if we wanted to calibrate a displacement (length) measuring system, we could use the appropriate primary standard, interferometry, which uses the wavelength of light as a length measuring device, since this is the same method as is used to define the fundamental unit of length, the metre. But we would be more likely to use a secondary standard, such as gauge blocks, which have themselves been calibrated using interferometry to a higher accuracy than that of the system we are calibrating, and which would be much easier to use.

RESPONSE TIME

Unless the quantity we want to measure has a steady value, or is changing only slowly with time, we must also consider the response time of the measurement system, to make sure that the display has reached its final value before the measured quantity has had time to alter appreciably.

Response time is determined by applying a *step input* (that is an instantaneous change of input, as in Fig. 1.19) to the system.

Fig. 1.19 Step input

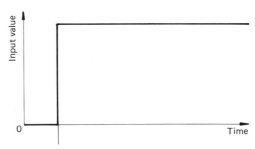

A *first-order system* (i.e. one whose response can be modelled by a first-order differential equation) responds as shown in Fig. 1.20, and the response time is the time it takes for the output to get within some specified percentage (say 5%) of its final value.

Fig. 1.20 First-order response

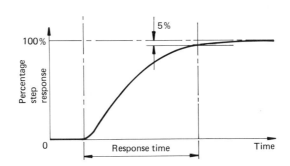

A *second-order system* (i.e. one whose response can be modelled by a second-order differential equation) responds as shown in Fig. 1.21; the response will *overshoot* by a percentage which depends on the *damping ratio* (this is explained in Chapter 11). The result is a *decaying oscillation*, and the response time in this case is the *settling time* — that is, the time it takes for the output to settle within some specified percentage (say ± 5%) of its final value.

Fig. 1.21 Underdamped second-order response

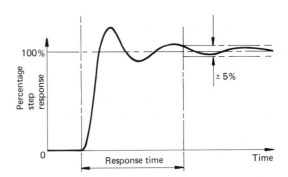

In some measurement systems there may be an unavoidable time delay between the instant of the step input and any visible change in the output. For example, in a temperature measurement system in which the heat has to conduct through an insulator to the sensing element, there will be a time delay between the temperature step and any detectable change in the signal from the sensor. Any such delay must be included in the response time.

RANGE, CAPACITY

The *range* of a measurement system is the difference between the maximum and minimum values which can be measured by it. Thus a thermometer graduated from $-20\,^{\circ}\text{C}$ to $110\,^{\circ}\text{C}$ has a range of $110-(-20) = 130\,\text{deg C}$, although its maximum reading is $110\,^{\circ}\text{C}$.

In some measurement systems, the maximum input which can be applied is called the *capacity* of the system. If we exceed the capacity (usually this is the f.s.d. value of the display) we are liable to upset the calibration, or, worse, to destroy the system altogether — it is very easy (and expensive) for instance to burn out the coil of a moving-coil meter or to burst the bulb of a glass thermometer by applying excessive inputs to such instruments.

RESOLUTION

The **resolution** of a measurement system is an indication of the 'fineness' of the measurements which can be made with it.

For example, using a steel rule to measure the diameter of a steel bar, I might record a measurement of, say, 32.0 mm. But with only average eyesight, I should have to admit that the only thing I could be sure of was that the measurement was not as big as 32.5 mm, and not as small as 31.5 mm. It could have been anywhere between those two values, so to me, the resolution of the steel rule — any steel rule — is 0.5 mm.

If I used a vernier caliper to take the measurement, I might measure the diameter as 32.2 mm. Since I could be sure that the reading was not as big as 32.3 mm, nor as small as 32.1 mm the resolution of this measurement system would be 0.1 mm.

Using a micrometer, I might read the diameter as 32.18 mm and be sure that the reading was not as big as 32.185 mm nor as small as 32.175 mm, so the resolution of this measurement system would be 0.005 mm.

These are all measurements obtained from *analogue* displays. An analogue display is one which does not give the measured value as a number, but indicates it as a position along a graduated scale (i.e. by *analogy*). Thus clocks (at least, those with hour and minute hands), moving-coil meters and in fact all pointer-and-scale instruments give analogue displays.

The alternative kind of display is one which shows the reading as a number, which is altered as the reading changes. This is a *digital* display (from *digits* meaning fingers — on which we count — hence a digit is one of the figures 0, 1, 2, ... 9). Obviously, when looking at a steady reading, we cannot tell how close the least significant digit (the right-hand digit) is to changing, so the resolution of a digital display is one (1) in the least significant digit — provided that the rest of the system has a resolution no worse than this.

For a comparison of analogue and digital systems, see p. 84.

Note that resolution does not mean the same thing as accuracy. A measurement system can have good resolution and yet have poor accuracy, because of, say, a large zero error. Such a system could still, however, be useful for comparative measurements. For example, a micrometer with large zero error could still be used to measure the ovality of a crankshaft journal accurately.

DRIFT

A measurement system can have a perfectly steady input and yet display an output which gradually changes as in Fig. 1.22. This effect is known as *drift*. It is apt to occur where the signal is in the form of an electric current or a voltage, and is usually caused by slight changes in resistance as the circuit warms up when switched on, or by changes in the gain of an amplifier. Strain gauge bridge circuits are sometimes subject to drift in this way. Where possible, such systems should have their power supplies switched on for an hour or two before measurements are to be taken, so that temperatures, (and hence resistance values and amplifier gains) can all become steady.

Fig. 1.22 Drift

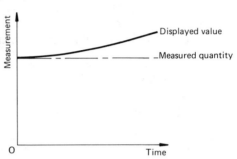

One method of overcoming drift in a d.c. voltage signal is to pass it through a *chopper*. This is a circuit which rapidly switches the signal on and off (i.e. chops it). The result is an a.c. voltage which alternates between the signal voltage and zero, at the switching frequency. This can then be amplified by an a.c. amplifier — a.c. amplifiers are much more stable than d.c. amplifiers.

DEAD ZONE

This is an effect which can occur when the signal is, for example, a rotation transmitted through gearing, or through a chain-drive. If there is too much clearance between the gear teeth (*backlash*) or too much slackness in the chain, the input may rotate through a small angle in either direction, before the slackness can be taken up and a change in output occur (see Fig. 1.23). The resulting error in output is particularly serious when the component forms part of an automatic control system.

Fig. 1.23 Dead zone

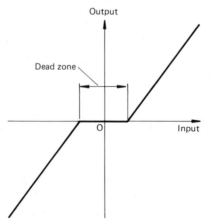

SUMMARY

Any measurement device can be considered as a system, and summarised by the *block diagram* of Fig. 1.24.

Fig. 1.24 The basic block diagram

The **transducer** converts a property which is difficult to measure directly into a form which is easier to measure. Its output is the **signal**.

The **signal conditioner** converts the signal from the transducer into a form suitable to operate the display unit.

A recording unit may be substituted for the display unit if a record of a time-varying measurement is required.

In more sophisticated systems the signal conditioner block can be subdivided into a series of blocks, each in its turn modifying the signal.

If the values of $\dfrac{\text{output}}{\text{input}}$ (the signal gain) of all the blocks are multiplied together, the result is the overall value of $\dfrac{\text{output}}{\text{input}}$ (i.e. the sensitivity of the system).

Common types of display unit are the Bourdon tube pressure gauge and the moving-coil meter, but these can also be treated as measurement systems in their own right. The moving-coil meter may be arranged to be current-operated (ammeter) or voltage-operated (voltmeter).

Accuracy is defined in terms of error.

Error is (indicated value) minus (true value).

Percentage error is $\dfrac{\text{error}}{\text{true value}} \times 100\%$

The **resolution** of a measurement system is the smallest *change* it can measure.

f.s.d. (full scale deflection) is the highest value on the scale of a scale-and-pointer display.

Zero error is a *constant* error which occurs throughout the range of the measurement system.

Tolerance is the allowable variation in a dimension.

The **reproducibility** or **repeatability** of a measurement system is determined by the overall variation in the displayed value when a series of repeated measurements of the same quantity is made within a short time, under fixed conditions.

The **stability** of a measurement system is determined by the overall variation in the displayed value when groups of repeatability tests are made, with long time intervals between the groups of tests.

The **constancy** of a measurement system is determined by the variation in the displayed value when the test conditions are varied within specified limits while a continuous steady quantity is provided as input.

Secular change is gradual change caused by ageing of materials.

Calibration is the determination of the relationship between the displayed value (the output) and the measured value (the input) over the full range of a measurement system.

Hysteresis is a variation in the output value for a given input value, which depends on the direction in which the input is changing. Its characteristic is a calibration graph in the form of an elongated loop.

A **primary standard** is a measurement produced from 'first principles', i.e. using a means by which the measurement units are defined.

A **secondary standard** is a measurement produced or measured by some other means which has itself been checked against a primary standard.

A **step input** is an instant change in an input value.

Response time is the time (including any time delay) taken by the output to reach and remain within a specified percentage of its final value when a step input is applied.

The **sensitivity** of a measurement system is calculated as a change in the reading of the display unit divided by the corresponding change in the measured quantity; i.e. the overall $\dfrac{\text{output}}{\text{input}}$.

The **range** of a measurement system is given by
(highest measurable value) minus (lowest measurable value).

The **capacity** of a measurement system is the highest measurable value.

Drift is a gradual change in the output of a measurement system for no change of input.

A **dead zone** is a range of input values, symmetrical about the point at which the input changes sign, for which there is no change in the output of the system.

EXERCISES ON CHAPTER 1

1   Explain briefly, in your own words, the meaning of the following terms used in engineering instrumentation:

(a) transducer;   (b) signal   conditioning;   (c) display.
Give an example of each.

2

Fig. E1.1

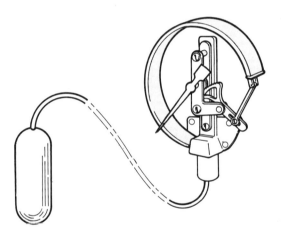

(a) A vapour-pressure type of temperature gauge consists of a steel bulb connected to a pressure gauge by a steel capillary tube, as shown in the diagram above. The bulb, capillary, and the Bourdon tube of the pressure gauge are filled with a vaporising liquid, so that the pressure in the system depends on the temperature of the liquid in the bulb. The scale of the pressure gauge is graduated in degrees Celsius. Draw the block diagram of the system, including blocks for the main components of the pressure gauge. There should be a block for each component which changes the signal into a different form or changes its magnitude.

(b) Indicate on the block diagram the transducer, the signal conditioning and the display.

(c) Indicate the amplifier in the system.

3   Fig. 5.15 on p. 122 shows a Bourdon tube pressure transducer with electrical output. The following data apply to the components:

> The link pin on the Bourdon tube deflects 3.5 mm for an increase in pressure of 1050 kPa.
>
> The link pin on the wiper is 6 mm from the wiper pivot.
>
> The resistance contact point on the wiper is 33 mm from the wiper pivot.
>
> The resistance element is 20 mm long, and the d.c. supply is 15 V.

(a) Draw a block diagram for the complete pressure measurement system, splitting the assembly shown in Fig. 5.15 into its component blocks, and adding a display device. Calculate the signal gain of each block.

(b) Determine the overall sensitivity of the pressure measurement system.

4   An instrumentation system to measure the bending moment on a wind-tunnel fan blade consists of:

(a) a bridge circuit of four active strain gauges, the output sensitivity of the bridge being $4\,\mu V/N\,m$;

(b) an operational amplifier to amplify the output of the bridge circuit;

(c) a power amplifier with a sensitivity of $20\,mA/V$;

(d) a pen recorder which requires $45\,mA$ for the full scale deflection of 30 mm.

For timing purposes, a square wave function generator also feeds into the power amplifier, having its output connected in parallel with that of the operational amplifier.

(a) Draw the block diagram of the complete system.

(b) Determine the gain of the operational amplifier if a bending moment of 2500 Nm is to give a pen deflection of 24 mm.

(c) Determine the amplitude of the output from the square wave generator, in volts, to superimpose a trace displacement of 5 mm amplitude on the trace from the pen recorder.

5   Fig. 5.7 on p. 116 shows a level-measuring system for a large storage tank.

The power supply is maintained at 12 V exactly, the depth of liquid to be measured ranges from 100 mm to 4600 mm, the drum is 200 mm diameter, and the potentiometer shaft can rotate through a maximum of 270°.

(a) Draw the block diagram, indicating the block(s) which correspond to (i) transducer, (ii) signal conditioning and (iii) display.

(b) Determine a suitable gear ratio to give a total rotation of the potentiometer shaft of between 250 and 270°.

(c) Hence determine the overall sensitivity of the system, assuming that a potentiometer shaft rotation from 0 to 270° corresponds to the range from 0 to 12 V on the voltmeter.

6   Each of the statements (a) to (e) which follow is taken from the specification of some measurement instrument. In each case explain briefly what the statement means. How would you check the truth of the first three statements?

(a) Sensitivity: 2 mV/cm;

(b) Repeatability: $\frac{1}{2}$% of span;

(c) Stability better than 0.2% of maximum span for six months;

(d) Accuracy (at 23 °C) ± 2% f.s.d.;

(e) Response time: 50 ms (10% to 90%).

7   The load indicated on the dial of a tensile testing machine is found to vary from the true value within the limits shown below:

± 100 N when scale is switched to read from 0 to 50 kN,
± 300 N when scale is switched to read from 0 to 100 kN,
± 500 N when scale is switched to read from 0 to 200 kN,
± 1500 N when scale is switched to read from 0 to 500 kN.

The maximum load the machine can apply is 500 kN.

Insert the numerical values in the following specification for the machine:

Range (kN)				
Accuracy (% f.s.d.)				

Capacity of the machine:

8   A spring intended for use as the transducer in a Post Office letter scale was tested and found to deflect 5 mm when a load of 4 N was applied to it. Select the correct value of signal gain for this transducer, from the values given below:

(a) 0.001 25 N/m,   (b) 0.001 25 m/N,   (c) 800 m/N,

(d) 20 N mm,   (e) 800 N/m,   (f) 0.8 N mm.

9   What adjustment would you make to eliminate a zero error in:

(a) a moving coil meter;   (b) a Bourdon tube pressure gauge?

10   Explain the construction and principle of action of any two of the following instruments or gauges, giving, in each case, a block diagram of the instrument as a self-contained measurement system:

(a) a Bourdon tube pressure gauge;

(b) a moving coil meter;

(c) a thermometer.

11   What is a transducer?

List ten of the physical properties for which transducers are available, and describe the main features and principles of two of these transducers.   (I.Q.A.)

12   An ammeter was calibrated against a primary standard producing known values of current. Readings were taken while the current was slowly increased from zero to full-scale deflection, and then while it was slowly decreased to zero again. The following results were obtained.

Ammeter reading (amps)	0	10	20	30	40	50	60	70	80	90	100
True current (amps)	0	11.3	22.0	32.4	42.5	52.4	62.3	72.0	81.2	89.7	97.5

Ammeter reading (amps)	100	90	80	70	60	50	40	30	20	10	0
True current (amps)	97.5	88.4	79.0	69.5	59.8	50.1	40.2	30.0	19.6	9.4	0

(a) Plot a graph of error against displayed value, as in Fig. 1.18 (note that *error = displayed value − true value*), and use this to plot a correction curve for the ammeter.

(b) Specify the accuracy of the ammeter as a percentage of f.s.d., when the correction curve is used.

(c) What is the effective accuracy of the ammeter when used with the correction curve (i) at half-scale and (ii) at quarter-scale deflection? Give your answer as a percentage error on the indicated value.

(d) What is the name of the effect which results in the graph forming a loop as in this example?   (I.Q.A.)

13   A displacement measuring system is provided with a meter scaled in 50 equal divisions, which acts as the display device. The system was calibrated by applying known displacements (D) to the system and noting the corresponding meter readings (M).

The following results were obtained:

D (mm)	5	6	7	8	9	10	11	12
M (scale divisions)	23.5	26	29	31	33	35	36.5	38

D (mm)	13	14	15	16	17	18	19	20
M (scale divisions)	39.5	41	42	43	44	45	46	47

(a) Draw the calibration graph for the system.

(b) Determine a suitable straight-line approximation to the calibration curve, and state it as an equation of the form $D = aM + b$ where a and b are numerical values to be determined from your straight line.

(c) State the maximum error involved in using your straight-line approximation over the range of displacements for which results were obtained.   (I.Q.A.)

14 Explain the meanings of the statements in the following extract from the specification of a pressure transducer:

> *Accuracy*
>
> Sensitivity $\leqslant 0.2\%$ of range.
> Linearity and hysteresis $\leqslant 1.0\%$ of range.
> Repeatability $\leqslant 0.1\%$ of range.

How would you check the last statement?

15 (a) Explain:
   (i) the difference between *stability* and *constancy* in a measurement system.
   (ii) the difference between a primary standard and a secondary standard for calibration purposes.

(b) By means of sketched graphs compare the response to a step input of (i) a first-order system and (ii) a second-order system. Show in each case how the response time is measured.

16 Explain what you understand by the term *resolution*. Give, as examples, two or three measurement systems which measure the same property with different resolution.

17 Explain, with the aid of sketched graphs, the meaning of the following terms as applied to instrumentation systems:

(a) dead zone;

(b) drift;

(c) hysteresis.

Chapter 2

Transducers

As we saw in the first chapter, a transducer is something which converts the property we want to measure into some other form easier to work on: easier to amplify, easier to transmit to another place, or more suitable to operate a display unit.

Almost anything can function as a transducer; the only essential is that it should produce an output change related to the input change by a fixed relationship.

THE SPRING

This is one of the simplest transducers. It converts force into a proportional displacement. so it can be thought of in block diagram form as shown in Fig. 2.1. Because it is entirely mechanical, nothing can go wrong with it, *provided it is not overloaded.* Overloading may introduce a permanent set in the spring, causing a change in the relationship between force and deflection, and hence altering its gain as a transducer.

Fig. 2.1 The block diagram representation of a spring

Fig. 2.2 shows various forms of spring used as transducers. The tension type (a) is preferable to the compression type (b), which, if badly proportioned, may buckle sideways under load. The hair-spring (c) we have already met in the moving-coil meter (p. 8) where it acts as a secondary transducer, following the original transducer and converting its output signal into yet another form. The spring beam shown at (d) is taken from the Hounsfield Tensometer (see p. 117), while the torsion spring (e) may be used where a very stiff, compact transducer is required, for example, in a weighbridge.

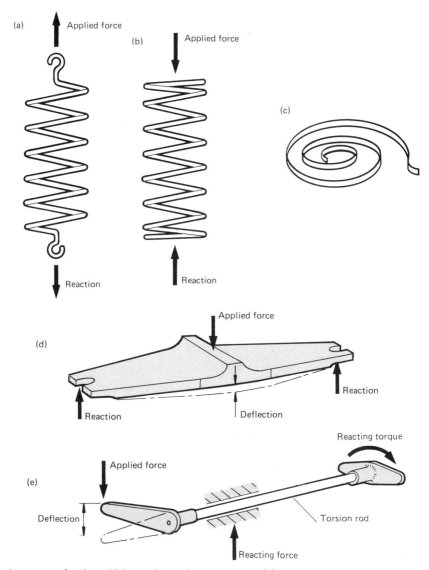

Fig. 2.2   Various types of spring which may be used as transducers: (a) tension spring,
(b) compression spring, (c) torsion spring (or hair spring), (d) spring beam, (e) torsion rod

RESISTANCE TRANSDUCERS

Resistance transducers convert the change in the property being
measured into a change of electrical resistance. Since a change in
electrical resistance can only be found by passing a current through
the resistance, resistance transducers *always* need an electrical
power supply. However, they have the advantage that their output
is already a voltage or current change, which makes the design of
the signal conditioning which follows them much more flexible
and much more compact. Two types of resistance transducer will
be considered: the strain gauge and the thermistor.

STRAIN GAUGES

When an electrical conductor is stretched, it increases in length and gets thinner. Both of these effects cause its electrical resistance to increase slightly. This is the principle of operation of the strain gauge.

There are several different kinds of strain gauge; the variations can be classified according to the 'family tree' in Fig. 2.3.

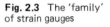
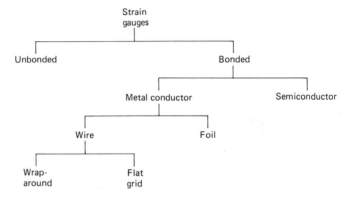

Fig. 2.3 The 'family' of strain gauges

An *unbonded strain gauge* consists of a continuous fine wire strung between two sets of insulating pegs. It is used to detect relative movement between two bodies. It is so delicate that it normally occurs only as a part of another transducer, in which case it is protected by the transducer casing. For example, some accelerometers contain unbonded strain gauges which convert the relative movement between the seismic mass and the body of the accelerometer into an electrical signal. Fig. 2.4 shows the arrangement of gauge wires in such an accelerometer.

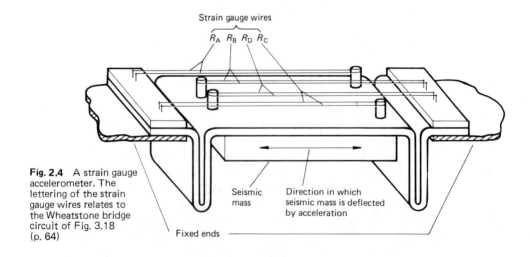

Fig. 2.4 A strain gauge accelerometer. The lettering of the strain gauge wires relates to the Wheatstone bridge circuit of Fig. 3.18 (p. 64)

Except for a very few such special purpose transducers, all strain gauges are *bonded* — that is, rigidly stuck with a suitable adhesive to a part of a structure or machine. This makes the conductor undergo exactly the same mechanical strain as the material to which it is stuck.

Bonding the gauge to the strained material makes it work for compressive strains as well, so that, just as tensile strain makes its resistance increase, compressive strain makes it decrease.

The three main types of bonded gauge are shown in Figs. 2.5, 2.6 and 2.7. Each type is sensitive to strain along axis Y-Y, and, as far as possible, insensitive to strain along the axis at right angles.

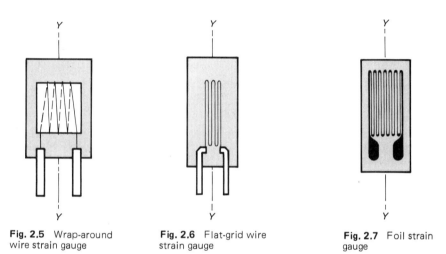

Fig. 2.5 Wrap-around wire strain gauge

Fig. 2.6 Flat-grid wire strain gauge

Fig. 2.7 Foil strain gauge

In the *wrap-around gauge*, the strain gauge wire is wound on to thin flat card; the winding is then sandwiched between two thin sheets of paper or plastic.

In the *flat-grid gauge*, the strain gauge wire is folded so that several lengths lie side by side, in the same plane. Like the wrap-around gauge, the wire is sandwiched between paper or plastic.

The wire gauge was the original form of strain gauge. They are still much in use, but tend to be superseded by the foil gauge, which has a better ratio of width to cross-sectional area for the conductor, giving better adhesion and heat dissipation.

The *foil gauge* has a zig-zag pattern of conductor stamped or etched out of thin metal foil and mounted on a thin plastic sheet base.

Semiconductor strain gauges are a recent addition to the strain gauge types available. The conductor is a crystal of germanium or silicon, treated with an impurity to make its resistance sensitive to strain. These gauges are about a hundred times as sensitive as ordinary strain gauges, so they can be used to measure very small strains, and they can be made small enough to measure strain virtually at a point, or in places where access is restricted, but at present they tend to be more expensive and less stable than ordinary gauges.

STRAIN CALCULATION

Mechanical strain (usually denoted by the Greek letter epsilon, ϵ) is calculated as

$$\epsilon = \frac{\text{Elongation}}{\text{Original length}}$$

The corresponding electrical property of a resistance, electrical strain, is

$$\frac{\text{Increase of resistance}}{\text{Original resistance}}$$

(Increase of a quantity is usually denoted by the Greek letter delta, δ. Thus electrical strain is $\delta R/R$.)

The electrical strain of a strain gauge is always directly proportional to the mechanical strain, so we can write

$$\frac{\delta R}{R} = k\epsilon$$

This is the essential equation for converting resistance change into mechanical strain.

The constant of proportionality, k, is called the *gauge factor* of the strain gauge. It is determined by the strain gauge manufacturer, from tests of samples of a particular gauge. It usually has a value of about 2.0, except in the case of semiconductor strain gauges, which have gauge factors in the range 100–300.

Gauge factors are the same for both tension and compression.

SELF-TEST QUESTION 1 (Solution on p. 230)

A strain gauge is affixed to a tie-bar of rectangular cross-section, as shown in Fig. 2.8. The resistance of the strain gauge is 120.27 Ω and its gauge factor is 2.1. The cross-section of the tie-bar is 25 mm $\times$ 6 mm, and the modulus of elasticity of its material is 200 GN/m².

Fig. 2.8 See question

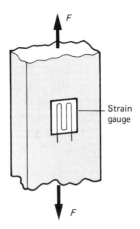

When a tensile load F is applied to the tie-bar, the resistance of the strain gauge is found to have changed to $120.42\,\Omega$. Determine:

(a) the strain in the material of the tie-bar,

(b) the stress in the material of the tie-bar,

(c) the load F.

SELF-TEST QUESTION 2 (Solution on p. 230)

An alternative loading on the tie-bar in the previous question induces a uniform *compressive* stress of $30\,\text{N/mm}^2$ on the cross-section. Determine the resistance of the strain gauge when the tie-bar is carrying this new stress.

BONDING TECHNIQUE

To adapt an old proverb slightly: 'a system is only as strong as its weakest component', and probably the weakest part of a strain gauge measurement system is the adhesion of the gauge to the metal under test. The electrical connections we can test and remake if necessary, but there is no way of determining whether a strain gauge is partly unstuck, and hence giving false readings. So bonding procedures have to be scrupulously followed — with no short cuts — if we are to have confidence in our strain gauge results.

The surface to which the gauge is to be applied must be cleaned down to the bare metal; any paint, rust or scale must be removed by file and emery until the metal shines smooth and bright. The area should then be degreased with a suitable solvent and finally neutralised with a weak detergent solution. Tissues or lint-free cloth should be used for this operation, wetting the surface and wiping off with clean tissues or cloth until the final tissue used is stain free, and the surface absolutely dry. Care must be taken not to wipe grease from a surrounding area on to the prepared area.

The strain gauge should be applied as soon as possible after this, before the surface can deteriorate. On no account must the cleaned surface or the underside of the strain gauge be touched by hand, as even the cleanest fingers leave a faint contamination. Strain gauges are normally supplied by the manufacturer ready for application, and when required for bonding, a strain gauge should be removed from the protective packaging using tweezers, and laid, bonding side downwards, on a previously degreased sheet of Perspex or similar plastic.

Modern cyanoacrylate adhesives are very convenient for strain gauge bonding. Their high-speed setting action, triggered by pressure, reduces setting time to seconds, but they must be handled with care, as they can bond fingers or eyelids together with equal facility. Any accidental contact with the adhesive should be washed away with water, instantly. Accidentally bonded fingers can be peeled apart using soap, hot water and a *blunt* parting tool such as a teaspoon, but it is better to avoid the need for such remedies. The following procedure should ensure that only the strain gauge gets stuck.

(1) Stick a piece of Sellotape over the strain gauge as it lies, bonding side downwards, on its Perspex sheet.

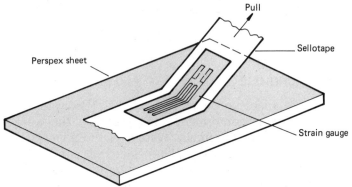

Fig. 2.9   Picking up the strain gauge by means of the Sellotape

Pull

Sellotape

Perspex sheet

Strain gauge

(2) Gently peel off the Sellotape and gauge, by pulling the tape upwards from one end, as shown in Fig. 2.9 taking care not to bend the gauge through too sharp an angle.

(3) Stick the Sellotape on to the already prepared site on the component to be tested, so that it holds the gauge against the metal in the position in which it is to be finally stuck.

(4) Gently peel back one end of the Sellotape, as in operation (2), and stick it back on the metal behind the other end, so that the strain gauge is suspended upside down, as in Fig. 2.10 ready to have its underside coated with adhesive.

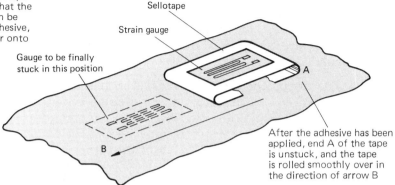

Fig. 2.10 Sellotape positioned so that the strain gauge can be coated with adhesive, then rolled over onto its final site

Sellotape

Strain gauge

Gauge to be finally stuck in this position

A

B

After the adhesive has been applied, end A of the tape is unstuck, and the tape is rolled smoothly over in the direction of arrow B

(5) Using the applicator nozzle of the tube of adhesive, spread the thinnest possible film of adhesive on the bonding face of the strain gauge. The gauge should look wet, but only just. Too much adhesive makes a poorer bond, and increases the risk of accidental contact with the skin.

(6) Reverse operation (4), by pressing the Sellotape and gauge smoothly back into its original position, with one stroke of even pressure from the fixed end of the Sellotape to the free end. Press through the Sellotape on to the gauge with finger or thumb for one minute, to make the adhesive set. After a further three minutes the Sellotape can be gently peeled off — but take care not to peel off the wire ends of the gauge with it.

Finally, after the connections have been soldered, the gauge can be protected and damp-proofed with a coat of air-drying varnish. For a more permanent installation, the gauge and its electrical connections may be encapsulated by spreading silicone rubber compound over them, so that they become embedded in a flexible coating which sets within 24 hours.

RESISTANCE TRANSDUCERS FOR TEMPERATURE MEASUREMENT

Most metals increase in electrical resistance as their temperature increases. This principle is made use of in the temperature measurement device known as the *resistance thermometer*. The change of resistance induced by a small change of temperature is very small, however, so to construct a measurement system with sufficient resolution, signal conditioning in the form of a Wheatstone bridge circuit (see Chapter 3) is a necessity. This makes the resistance thermometer a rather delicate, rather cumbersome system, though it is widely used for the precise measurement of temperature, especially of high temperatures.

The Thermistor

What we usually want is a temperature transducer which could be less precise, but with a much greater sensitivity, so that it could feed directly into a display unit without the need for signal conditioning.

The thermistor is such a transducer. It is a kind of semiconductor — that is, a conductor whose resistance varies as the relevant operating condition (in this case, temperature) is varied. The material of a thermistor is usually a metallic oxide, chosen so that its resistance, R, follows a law of the form

$$R = Ae^{B/T}$$

where A and B are constants and T is absolute temperature. This gives a very big *decrease* in resistance for a comparatively small increase in temperature. Fig. 2.11 shows the characteristic temperature/resistance curve of a typical thermistor, together with the corresponding curve for copper. In order to make the two curves comparable, the ratio $\dfrac{R}{R_0}$ (i.e. the ratio of the actual resistance to the resistance at $0\,^\circ$C) has been plotted, instead of R.

It can be seen from Fig. 2.11 that the thermistor can be used as a temperature transducer over only a comparatively small range of temperature. A typical application is the water temperature measurement system on a motor car. Here we do *not* want precise measurement; all we need is an indication of one of three states — cold/normal/hot — to be provided by a cheap, robust system. The result is the circuit shown in Fig. 2.12. The current meter will almost certainly *not* be a moving-coil meter — a pointer carried by a bimetallic strip heated by the current is the usual form of display device.

Fig. 2.11 Relative resistance curve for copper and for a typical thermistor

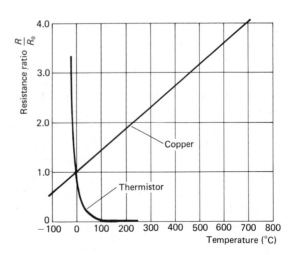

In industry, thermistors are used for much more precise temperature measurement than this, in situations where their maximum useful temperature limit of about 300 °C will not be exceeded. They have the advantage of high sensitivity, and as the bead of semiconductor material can be made very small, they can measure temperature virtually at a point, with very fast response times.

Fig. 2.12 (a) Circuit diagram for motor car water temperature indicator system. (b) Corresponding block diagram

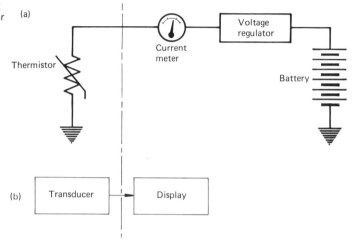

SELF-TEST QUESTION 3 (LABWORK) (Solution on p. 231)

In a test to determine the characteristics of a thermistor used as a motor-car water-temperature transducer, the thermistor was suspended in a mixture of ice and salt which was gradually raised to boiling point and then allowed to cool. At intervals, the temperature of the mixture was taken with a thermometer, and the resistance of the thermistor was measured on a digital multimeter. The following results were recorded.

Temperature (°C)	5	3.5	17.5	36	55	76
Resistance (Ω)	3260	1831	765	263	97.7	36.2
Temperature (°C)	85	100	97	66	43	21
Resistance (Ω)	24.3	13.3	14.2	55.6	171.4	581

(a) Plot the calibration graph of the thermistor.

(b) Determine the law of the thermistor.

PIEZOELECTRIC TRANSDUCERS

A piezoelectric material is a special kind of crystalline electrical insulator which, when compressed along one of its axes, acquires an electrostatic charge on opposite faces of the material. The compressive force is usually applied through metal electrodes (as in Fig. 2.13) which serve to spread the load evenly over the material and to make the electrical connections to it.

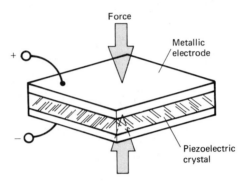

Fig. 2.13 The piezo-electric effect

A piezoelectric transducer can thus be thought of as a capacitor, but a special kind of capacitor — one which acquires its charge when a compressive force is applied to it. Its input is force and its output is charge.

The piezoelectric material may be quartz or Rochelle salt, which occur naturally, or a ceramic such as lead zirconate-titanate.

By the law of conservation of energy, the electrical energy of the charge must come from the work done in compressing the crystal. Only a very small amount of work is done, as piezoelectric materials have some of the highest moduli of elasticity of all solids; hence they deflect very little when compressed. However, the capacitance of a piezoelectric transducer is very small indeed, so a small electric charge can produce quite a high voltage $\left(\text{remember: voltage} = \dfrac{\text{charge}}{\text{capacitance}} \right)$. For instance the voltage is high enough to flash over into a spark when a piezoelectric crystal is used as the ignition source in gas lighters and cigarette lighters.

Once the crystal becomes compressed by a force, the work has been done: the charge on the crystal has been created. No more charge will be produced unless the force is increased.

We have to find a way of measuring this charge without letting it leak away through the measuring equipment. This seems an impossible task, because we can only measure the charge by measuring the voltage it produces between the faces of the crystal, and our voltage measuring equipment cannot have an infinitely high resistance. A moving-coil voltmeter, for instance, applied to the crystal, would absolutely short-circuit it; the pointer would never leave the zero mark on the scale.

In fact, it is possible to measure the charge on the crystal while drawing almost no current at all from it, by connecting the piezoelectric transducer to a special kind of amplifier called a *charge amplifier*. This is described in Chapter 3, p. 77.

Even a charge amplifier draws a very slight current from the piezoelectric crystal, so the voltage between the faces of the crystal falls to zero in time. Fig. 2.14 shows (b) the output voltage of a piezoelectric transducer when (a) a step input of force is applied. Obviously it is no use applying a steady load to the transducer and expecting it to be still giving the same output a day or two later, but if the fall in voltage is slow enough, we might be able to use it to measure steady loads if we measure them within a few seconds of applying them. We can determine whether this is possible from the *time constant* of the transducer-and-amplifier combination.

Fig. 2.14 Step response of a piezoelectric transducer: (a) input; (b) output

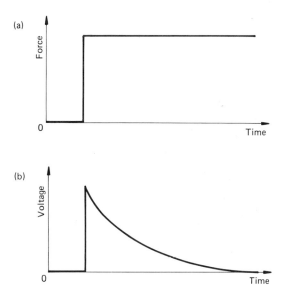

The response curve in Fig. 2.14(b) is an exponential curve in which the output voltage is given by $v = V e^{-t/\tau}$ where V is the voltage at the beginning of the step response, t is the time (in seconds) after the input step, and the Greek letter τ (pronounced 'taw') is the time constant of the system. (e is, of course, the mathematical constant 2.718...) The time constant is the time it takes (in seconds) for the output to fall to 36.8% of its original value. Why 36.8%? Because when $t = \tau$

$$V e^{-t/\tau} = V e^{-1} = 0.368\,V$$

SELF-TEST QUESTION 4 (Solution on p. 233)

A piezoelectric force transducer was connected to a charge amplifier. A step input of force was applied to the transducer and the output voltage of the charge amplifier was read at intervals after the step was applied. The following results were obtained:

Time in seconds since step was applied	0	2	4	6	8	10	15	20	25
Output voltage of charge amplifier	6.0	4.2	2.9	2.1	1.4	1.0	0.4	0.2	0.1

The output voltage of the charge amplifier was zero just before the step was applied.

Determine the time constant of the system.

SELF-TEST QUESTION 5 (Solution on p. 234)

Use your calculator to find what percentage of the original output value is left at the following times after the step input:

$$0.01\tau,\ 0.02\tau,\ 0.05\tau,\ 0.1\tau,\ \tau,\ 2\tau,\ 3\tau,\ 4\tau \text{ and } 5\tau$$

From the solution of the self-test question (p. 234) we can see that for time intervals up to about 0.1τ we are losing approximately 1% of the output of the crystal in each hundredth of the time constant after the step input was applied (i.e. the output curve and tangent PQ in Fig. S.3 (p. 233), approximately coincide over this interval). As the time interval increases, the rate of loss of output slows down, but by the time three time constants have elapsed we are down to 5% of original output of the crystal, and after five time constants the charge on the crystal has virtually all leaked away.

So if we know the time constant of a piezoelectric transducer-and-charge amplifier force measurement system we shall know what time is available for measuring a steady load before the error becomes too great, and we can even correct our readings to allow for the leakage of charge which has occurred between the application of the force and its measurement.

Piezoelectric transducers are mainly used to measure alternating loads, however, such as forces due to vibration. When the charge amplifier is switched to this kind of measurement, the system may have a shorter time constant, so that measurements made at very low frequencies may be inaccurate. In this case, the manufacturer of the charge amplifier would specify a low frequency limit for dynamic measurements and at frequencies below this, the charge amplifier would be switched to *static* (steady load) measurement. (This statement will become clearer when you have read the section on charge amplifiers on p. 79.)

THE LINEAR VARIABLE DIFFERENTIAL TRANSFORMER (LVDT)

This is a transducer which converts linear displacement into an electrical signal in the form of an alternating current, the displacement information being given by the amplitude and the in-phase or anti-phase state of the a.c. Fig. 2.15 shows the essential details of the transducer.

Fig. 2.15 Diagram of the LVDT transducer

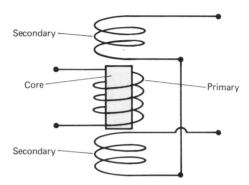

An alternating current of constant amplitude and frequency is passed continuously through the primary coil, and the alternating magnetic field thus set up induces alternating currents of the same frequency in the secondary coils. These secondaries are identical, and are symmetrically placed about the primary so that all three coils are on the same axis. The electromagnetic coupling between the primary and the secondaries is greatly increased by a cylindrical ferromagnetic (e.g. iron) core.

The displacements to be measured are applied to the core by a non-magnetic (e.g. brass) rod which carries the core. These displacements move it along the axis in one direction or the other. When the core is in its mid-position, the secondaries pick up equal a.c. voltages. The secondaries are so connected that these voltages oppose each other, so that the output with the core in mid-position is zero. With the core in any other position, the output is an a.c. whose amplitude is proportional to the displacement from the mid-position, and which is in phase with, or in opposite phase to the current through the primary, depending on whether the core is on one side or the other of the mid-position.

This output a.c. is passed through a phase-sensitive detector which rectifies the a.c. and produces a positive- or negative-voltage d.c. depending on the in-phase or anti-phase state of the input to the detector. This d.c. can then operate a display device such as a centre-zero moving-coil meter (i.e. one in which the pointer rests in the middle of its range when there is no input).

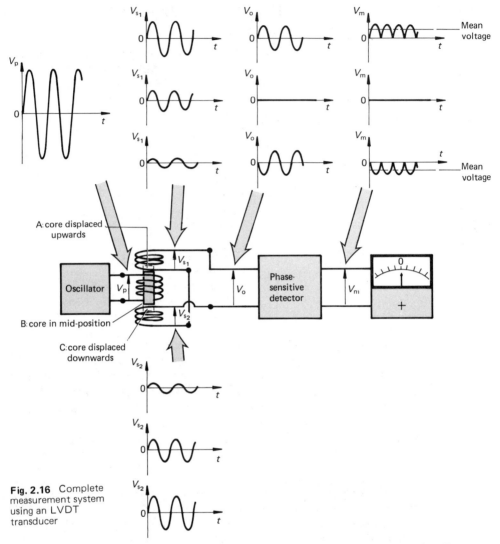

Fig. 2.16 Complete measurement system using an LVDT transducer

Fig. 2.16 shows the circuit diagram of such a system. The diagram also shows typical voltage/time graphs at various points in the circuit. These voltages are:

v_p a.c. supply to the primary

v_{s_1} voltage induced in upper secondary

v_{s_2} voltage induced in lower secondary

v_o output of the transducer $(= v_{s_1} - v_{s_2})$

v_m output from phase-sensitive detector

For v_{s_1}, v_{s_2}, v_o, and v_m the upper, middle and lower graphs correspond to core positions A, B and C respectively.

Fig. 2.17 is the corresponding block diagram.

Fig. 2.17 Block
diagram

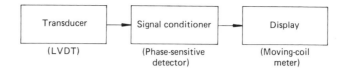

Transducer	Signal conditioner	Display
(LVDT)	(Phase-sensitive detector)	(Moving-coil meter)

The LVDT measurement system can be extremely sensitive, a resolution down to 0.05 mm being quite normal, though the range of displacement which can be measured is limited to a few centimetres. The output of the transducer is completely unaffected by any sideways movement of the core, since this does not affect the magnetic coupling between the coils.

The LVDT can be used as a secondary transducer in systems used for the measurement of (e.g.) force, or pressure, or acceleration, in which the primary transducer (a spring, or hydraulic bellows, or a spring-mounted mass) converts the force or pressure or acceleration into a proportionate displacement which in turn is applied to the core of the LVDT.

LIGHT-SENSITIVE TRANSDUCERS

The main types of light-sensitive transducer at present available are the *photoconductive cell*, and the *photovoltaic cell*. In deciding which type to use in a particular measurement system, we need to consider (a) the kind of power supply it would require, (b) its response time, and (c) its spectral response (this term is explained below).

THE NATURE OF LIGHT

Light is electromagnetic radiation (i.e. radio waves) of very short wavelength — of the order of half a thousandth of a millimetre. The wavelength of light determines its colour. Visible light ranges in wavelength from about 400 nm (violet) to about 700 nm (red). (nm is the abbreviation for nanometres; 1 nm is $1/10^9$ of a metre.) White light is a combination of all the colours of the spectrum, and can be considered as having a mean wavelength of 500 nm. However, transducers are available which can 'see' longer wavelengths than 700 nm (i.e. *infrared* light) or shorter wavelengths than 400 nm (i.e. *ultraviolet* light). In choosing light-sensitive transducers, therefore, we need to know their *spectral response* — that is, their sensitivity to light of various wavelengths.

THE PHOTOCONDUCTIVE CELL

This is a semiconductor whose electrical resistance depends on the intensity of illumination reaching the semiconductor material. It is also called a *photoresistive cell* or a *light-dependent resistor*. To use it as a transducer, a constant voltage must be applied to it. Its output signal is then a current which varies with the intensity of illumination.

Fig. 2.18 Photo-
conductive cell, about
three times full size;
sectioned to show
construction: (a) top
view, (b) side view

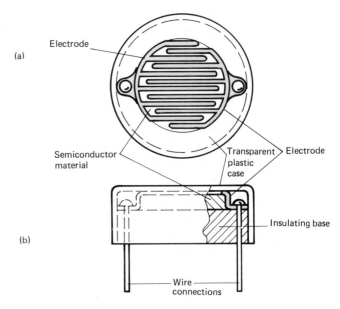

(a)

Electrode

Semiconductor
material

Transparent
plastic
case

Electrode

Insulating base

(b)

Wire
connections

Fig. 2.18 shows the construction of a photoconductive cell, and
Fig. 2.19 shows a simple circuit in which the current displayed by
the meter is a function of the illumination reaching the cell.
Photoconductive cells are the most sensitive of the light-sensitive
transducers: a typical cell varies in resistance from about $10\,\mathrm{M\Omega}$,
when the cell is in complete darkness, down to less than $100\,\Omega$
when the cell is under strong illumination. There is usually no
need of an amplifier, therefore, to boost its output signal.

Fig. 2.20 compares the spectral response of the two most commonly
used photoconductive materials with the spectral response of the
human eye. It can be seen that cadmium sulphide has a spectral
response similar to that of the eye, while cadmium selenide has a
substantial response in the infrared region. Cadmium sulphide
cells are therefore used for white-light and photographic equip-

Fig. 2.19 Simple light-
measurement system
using a photoconductive
cell as the transducer

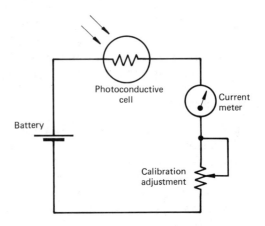

Photoconductive
cell

Current
meter

Battery

Calibration
adjustment

ment applications, and cadmium selenide for infrared measurement or detection systems. The output of an ordinary electric lamp consists of far more infrared than visible light (its output spectrum is included in Fig. 2.20), so where an electric lamp is the light source, the cadmium selenide cell is the more suitable, especially if the lamp runs at low power.

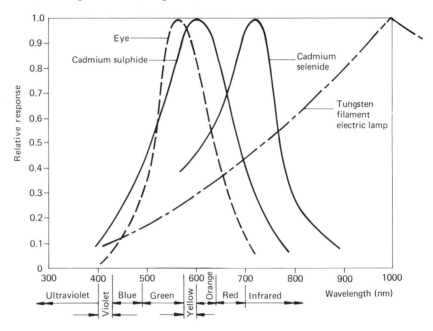

Fig. 2.20 Spectral response of the semiconductor materials cadmium sulphide and cadmium selenide compared with that of the human eye and with the output spectrum of an electric lamp

Photoconductive cells are the slowest of the light-sensitive transducers: resistance rise times of 0.075 seconds and fall times of 0.35 seconds are typical. They are therefore suitable for measuring only steady or slowly varying levels of illumination.

THE PHOTOVOLTAIC CELL (THE PHOTOJUNCTION CELL)

Unlike the photoconductive cell, which uses metal electrodes plated on to a single semiconductor material, the photovoltaic cell is activated by the illumination of the junction between two kinds of semiconductor material: an n-type material and a p-type material. n-type material contains a slight excess of electrons, and p-type a slight deficiency of electrons (or in the language of transistor technology, an excess of holes).

Illumination of the junction causes electrons to be released from the atoms of the p-type material. These electrons migrate across the junction into the n-type material — thus a current flows. They leave holes in the p-type material, which migrate in the opposite

Fig. 2.21 The photo-
voltaic cell as a
generator (corresponds
to sections ABC and
DEF of Fig. 2.22)

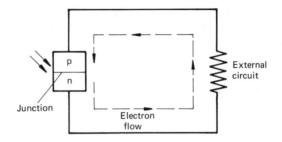

direction until they are filled by electrons entering from the
external circuit. Thus, as shown in Fig. 2.21, the p–n junction
acts as an electricity generator, the p material forming the positive
terminal of the generator, and the n material the negative.

The p–n junction forms a semiconductor diode, and when the
junction is kept in darkness, it behaves as an ordinary solid-state
diode, giving curve (i) of Fig. 2.22.

Fig. 2.22 Voltage-
current characteristics
of a photovoltaic cell

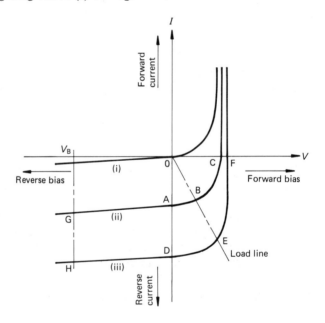

In Fig. 2.22, *forward bias* means the connection of the p material
to the positive of an external voltage source, and the n material to
the negative. *Forward current* means a flow of electrons through
the external circuit from the p material to the n material — hence
in the opposite direction to that shown in Fig. 2.21.

When the p–n junction is illuminated, curve (i) becomes shifted
downwards, giving curves (ii) and (iii). The amount of downward
shift is proportional to the intensity of illumination of the
junction; thus the illumination giving curve (iii) is twice that which
gives curve (ii).

The photovoltaic cell is made in two main forms, depending on the purpose for which it is to be used. One form is the *solar cell*, used to convert light into electrical power. Solar cells are made with p-n junction areas as large as possible, to absorb as much light as possible. They function on parts ABC or DEF of the curves. At points A and D there is current flowing but no voltage applied by the external circuit — that is, the cell is short-circuited. At points C and F there is voltage generated but no current flows — that is, the cell is on open circuit. Typical operating points are B and E which lie on a load line whose gradient, by Ohm's law, is $-1/R$, where R is the resistance of the external circuit.

The other form of photovoltaic cell is the *photodiode*, which is used where a light-sensitive transducer is required to have a fast, linear response to intensity of illumination. When it is used in this way, an external voltage V_B is applied to reverse-bias the p-n junction; this gives operating points G and H at which a reverse current flows, directly proportional to the intensity of illumination of the p-n junction.

Photodiodes are usually very compact units in which the p-n junction is very small indeed, so the signal current also is very small, and always requires amplification. Some photodiodes are obtainable with an already-coupled integrated circuit amplifier formed on the same chip of silicon as forms the p and n material of the photodiode.

Silicon and germanium are the two main materials used for photodiodes. Either can be turned into both p-type and n-type material by the introduction of traces of suitable impurities. Both elements have spectral responses compatible with that of the human eye, but the spectral response of germanium is broader, extending well into the long-wavelength infrared region.

The speed of response of a photodiode is very much better than that of a photoconductor; response times are of the order of microseconds ($1\,\mu s = 1/10^6$ seconds), instead of tenths of a second.

EXERCISES ON CHAPTER 2

1   Sketch and describe a measurement system which uses a spring as a transducer.

2   Draw a labelled sketch to show the construction of a bonded resistance strain gauge. State typical materials used, and the precautions to be used in fixing a gauge.

3   A strain gauge was used to measure the strain in a steel rod 6.5 m long, which was to be loaded in tension, the axis of the strain-gauge being parallel with the axis of the rod. When the tensile

SUMMARY OF LIGHT-SENSITIVE TRANSDUCERS

Transducer	Spectral response	Power supply required	Speed of response	Construction	Remarks
Photoconductive cell	*Cadmium sulphide:* approximately as human eye. *Cadmium selenide:* yellow to infrared (hence suitable for tungsten or infrared source)	Small DC supply, e.g. 1.5 V to drive current through cell	Slow: typical values 0.075 s rise time; 0.35 s fall time	Metal electrodes plated on to a single semiconductor material (Fig. 2.18)	Typically 10 MΩ (dark) to 100 Ω in bright light
Photovoltaic cell (Photojunction cell)	Silicon-based has response as human eye. Germanium based extends well into long infrared region	None for solar cell. A few volts biasing voltage for photodiode, but output signal needs amplifier	Of the order of 1 μs	p–n material illuminated at junction (Fig. 2.21) Photodiode very compact, or solar cell large area	Silicon or germanium

load was applied, the resistance of the strain gauge changed from 600.02 Ω to 600.69 Ω. The gauge factor of the strain gauge was 1.98. Determine:

(a) the strain in the rod;

(b) the extension of the rod.

4   To check the gauge factor of a batch of strain gauges, a sample gauge was bonded to an aluminium alloy rod. A tensile load was then applied to the rod while the change in resistance of the gauge and the extension of a 50 mm gauged length of the rod were noted.

The original resistance of the strain gauge was 199.73 Ω; this increased to 200.81 Ω when the rod was loaded. The corresponding extension of the gauged length was 0.128 mm. Determine the gauge factor.

5   (a) What is an electrical resistance strain gauge? How is it used to measure strain? Briefly describe the two main types commonly available.

(b) The following data were obtained from a tensile test on a strain-gauged rod.

Original resistance of gauge:	500.32 Ω
Final resistance of gauge:	501.46 Ω
Gauge factor:	2.04
Modulus of elasticity of the rod material:	200 GN/m^2
Diameter of rod:	14 mm

Determine for the rod:
(i)   the tensile strain;
(ii)   the tensile stress;
(iii) the tensile load.

6   (a) What is a thermistor?

(b) Sketch the output/input graph of a thermistor.

(c) Draw an electrical circuit diagram for a measurement system incorporating a thermistor and state a typical application for which this type of measurement system would be used.

7   The relationship between resistance and temperature for a thermistor is given by $R = A\,e^{B/T}$, the characteristic temperature, B, being 3050 K. If its resistance at 25 °C is 1680 Ω, determine its resistance:

(a) at 0 °C   and   (b) at 300 °C.

8   In a test to determine the characteristic temperature of a thermistor, the following results were recorded:

Temperature (°C)	21	30	40	50	60	70	80	90	100
Resistance (kΩ)	1117	680	449	278	174	113	75.5	51.0	35.3

Plot ln R against $1/T$, and use two points on the straight line to determine the values of the constants A and B in the equation for the resistance of a thermistor.

9 (a) What is the principle on which a piezoelectric transducer works?

(b) In what units is the sensitivity of a piezoelectric force transducer expressed?

(c) Sketch a voltage–time graph showing the response of a piezoelectric force transducer to a step input of force.

10 The following results were obtained when a step input of force was applied to a piezoelectric force transducer connected to a charge amplifier:

Time since application of force (seconds)	0	5	10	15	20	30	45
Output voltage (V)	9	6.3	4.5	3.1	2.2	1.1	0.4

The output voltage eventually reverted to zero. Determine the time constant of the system.

11 A step input of force is applied to a piezoelectric force measurement system which has a time constant of 20 seconds. Choose, from the list below, the nearest value to:

(i) the time it takes for the output to fall to 95% of its original value.

(ii) the time it takes for the output to fall to 5% of its original value.

(a)	(b)	(c)	(d)	(e)	(f)	(g)	(h)	(i)	(j)
0.1 s	0.5 s	1 s	5 s	19 s	20 s	40 s	60 s	80 s	100 s

12 When a thermometer is plunged into a liquid at a higher temperature, the temperature–time graph of its readings is an exponential curve, like the voltage–time graph for charge leakage from a piezoelectric transducer, but inverted. Thus the table of multiples of time constant on p. 234 applies in this case also.

(a) The graph shown in Fig. E2.1 was obtained when a resistance thermometer was suddenly transferred from water at 0 °C to water at 100 °C. Choose, from the following, the best estimate of the time constant of the thermometer:

(i) 1 s   (ii) 2 s   (iii) 4 s   (iv) 8 s   (v) 12 s   (vi) 16 s.

Fig. E2.1

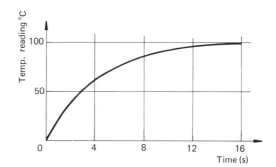

(b) If, when the thermometer is reading 100 °C, it is suddenly transferred to water at 50 °C, in which of the following time intervals will it indicate 55 °C?
(i) 0–1 s   (ii) 1–2 s   (iii) 2–4 s   (iv) 4–8 s   (v) 8–12 s
(vi) 12–16 s.

(c) If this thermometer is to give readings which are within 1 % of any step change of input, which of the following time intervals is the least which must be allowed?
(i) 4 s   (ii) 8 s   (iii) 12 s   (iv) 16 s   (v) 20 s   (vi) 24 s.

13 (a) Explain the principle of a differential transformer (LVDT) type of transducer.

(b) Show with the aid of a diagram how this type of transducer could be adapted to measure *one* of the following:
(i) pressure;   (ii) force;   (iii) acceleration.

(c) Briefly describe an alternative method of producing a voltage proportional to displacement.

14 Explain, with the aid of a sketch showing the magnetic flux paths, the operation of a linear variable differential transformer.

The displacement of a machine member from a datum position is sensed by a linear variable differential transformer. Give a circuit diagram and explain how the displacement may be indicated on a centre-zero moving-coil instrument.

15 Arrange ultraviolet, infrared and the colours of the visible spectrum in order of wavelength. Indicate on your list the approximate values of the wavelengths at which ultraviolet and infrared change into visible colours.

16 (a) Sketch the construction of a photoconductive cell.

(b) Draw a simple circuit for a light measurement system, using a photoconductive cell as transducer.

(c) The most commonly used semiconductor materials for photo-conductive cells are cadmium sulphide and cadmium selenide. Which material would you specify for:
(i)   a light meter for use by photographers;
(ii) an infrared detection system?
In each case give a reason for your choice.

17   (a) Give the names of the two types of photovoltaic cell, explain how each may be distinguished from the other, and state the main purpose of each type.

(b) Sketch the family of current–voltage curves which illustrate the relationship between voltage and current for a photovoltaic cell. Indicate on your sketch the operating region for each type.

(c) Explain what is meant by *reverse bias* and *reverse current* in the context of your answer to part (b).

18   Select from items (i) to (iii) below, the most suitable light-sensitive transducer for the following purposes. Choose also from items (iv) to (vii) the most suitable power supply for each.

(a) Conversion of light into electrical power.

(b) Very compact transducer required for high-speed (1 μs rise time) conversion of infrared light signal into electrical signal.

(c) Output of transducer to feed directly into display device without the use of an amplifier. Must be fairly compact. Speed unimportant.

Choose from: (i) solar cell, (ii) photoconductive cell; (iii) photo diode.
Power supplies: (iv) 15 V d.c.; (v) 100 V d.c.; (vi) 1.5 V d.c. (vii) none.

Chapter 3

Signal Conditioners

AMPLIFIERS

The word 'amplify' means 'to increase', so an amplifier is a form of signal conditioner which *increases* the signal in some way, without changing its nature, giving a mechanical or electrical output larger than its input.

MECHANICAL AMPLIFIERS

Mechanical amplifiers are *passive* devices, which are used to amplify displacements or rotations. They do not have any external power supply, other than the power put into them by the input signal. A suitable mechanical amplifier for linear displacements is the lever; for rotational displacement amplification, a pair of gears may be used.

Figs. 3.1 and 3.2 show the two possible forms of lever amplification. In both cases

$$\text{Gain} = \frac{\text{Output displacement}}{\text{Input displacement}}$$

$$= \frac{\text{Radius from pivot to output link}}{\text{Radius from pivot to input link}}$$

i.e. the gain is $\dfrac{b}{a}$ in Fig. 3.1 and also in Fig. 3.2.

Fig. 3.3 shows a simple gear train. Its gain is given by

$$\text{Gain} = \frac{\text{Output revolutions}}{\text{Input revolutions}}$$

$$= \frac{\text{Number of teeth on input gear}}{\text{Number of teeth on output gear}} \text{ *}$$

*Note the inversion: input over output.

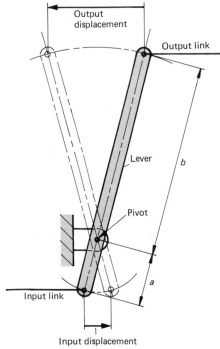

Fig. 3.1 Inverting
amplifier of displacement

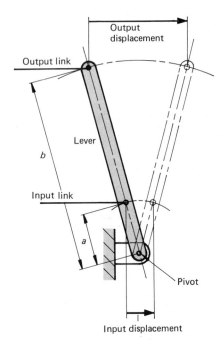

Fig. 3.2 Non-inverting
amplifier of displacement

Thus for the simple gear train shown in Fig. 3.3, because the input gear, A, has 12 teeth and the output gear, B, 9 teeth the gain of this mechanical amplifier must be

$$\text{Gain} = \frac{12}{9} = 1.333$$

In the case of compound lever systems (i.e. the output of one lever linked to the input of the next) we get the overall gain by multiplying together the gains of the separate levers.

Fig. 3.3 Simple gear
train

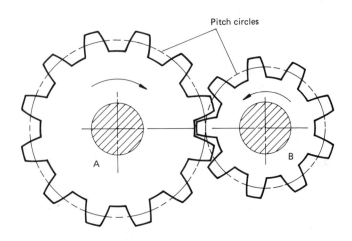

Similarly, in the case of compound gear trains (i.e. the output gear of one pair fixed to the input gear of the next pair), we get the overall gain by multiplying together the gains of the individual pairs of gears in mesh.

ELECTRONIC AMPLIFIERS

We use electronic amplifiers every day, whenever we switch on a radio or television set, or play a record player or cassette recorder. These all have amplifiers of various kinds to increase the signal from the tiny alternating voltage induced at the input, into the immensely greater current or voltage necessary to operate the output device.

Electronic amplifiers can be divided into two main types: *voltage amplifiers* and *power amplifiers*.

A voltage amplifier is one which increases the amplitude of the voltage applied to its input, as shown in Fig. 3.4. However, the current which can be taken from the output terminals of a voltage amplifier is practically negligible, so, if we want the output to do some physical work, such as operating the speech coil of a loud-speaker to vibrate the air, or rotating the coil of a moving-coil

Fig. 3.4 Amplification of an electric AC signal

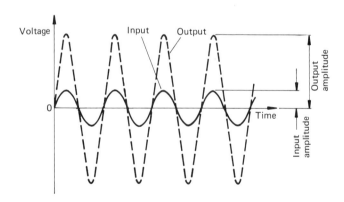

meter, we must feed the output of the voltage amplifier into the input of a power amplifier. A power amplifier gives little or no voltage amplification — in fact the output voltage amplitude may well be less than that of the input — but associated with that output voltage will be a large output current — so a power ampli-fier is basically a current amplifier.

In block diagrams of electrical systems, an electronic amplifier is often indicated by a triangle, as shown in Fig. 3.5, so a voltage amplifier feeding into a power amplifier would be as shown in Fig. 3.6.

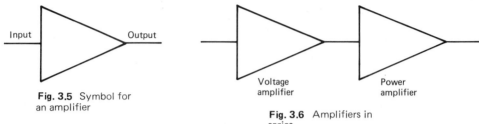

Fig. 3.5 Symbol for an amplifier

Fig. 3.6 Amplifiers in series

Nowadays, voltage amplifiers are usually bought complete as integrated circuits, in which a microscopic network of transistors, capacitors and resistors making up the amplifier is formed on a chip of silicon a few millimetres square. Fig. 3.7 shows approximately the actual size of a typical integrated-circuit amplifier and Fig. 3.8 shows how the internal connections to the silicon chip are made.

Fig. 3.7 Integrated-circuit amplifier. *Source:* RS Components Ltd, London

Fig. 3.8 Cutaway view of integrated-circuit amplifier showing connections to silicon chip. *Source:* RS Components Ltd, London

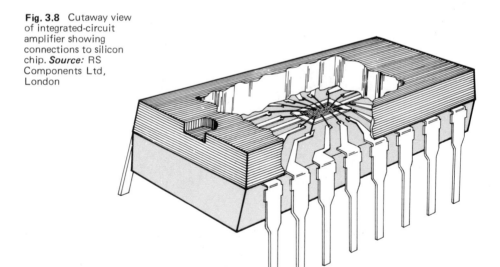

Integrated-circuit voltage amplifiers of the type shown in Fig. 3.7 are usually *operational amplifiers* — so called because they were originally devised to perform arithmetic operations (adding, subtracting, etc.) in analogue computers. Fig. 3.9 shows the circuit diagram symbol for an operational amplifier.

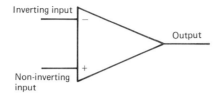

Fig. 3.9 Symbol for an operational amplifier

This type of amplifier amplifies only the voltage *difference* between the two inputs, so it may also be called a *differential amplifier*. The signal is usually fed into the inverting input, producing a voltage difference relative to the non-inverting input, and an amplified version of the signal appears at the output, with opposite polarity (opposite phase).

If, instead, the signal is fed into the non-inverting input, the output will have the same polarity as the input (i.e. it will be in phase with it). Operational amplifiers have the advantage that if an unwanted signal or 'noise' is picked up from nearby wiring, it tends to be picked up equally by both inputs, and so it is less likely to be amplified.

One important feature of electronic amplifiers, not made clear in the neat little diagrams of Figs. 3.5, 3.6 and 3.9, is this: *every electronic amplifier must be supplied with a constant voltage from a d.c. power supply*. Thus, for example, the operational amplifier of Fig. 3.9 should really be shown as in Fig. 3.10, though the power supply connections are usually omitted, along with other details, to avoid complication.

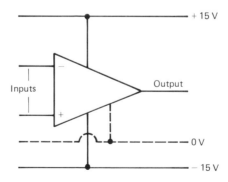

Fig. 3.10 The power supply connections to an operational amplifier

It may help in understanding the behaviour of an electronic amplifier if we think of it as being like a water tap. The input signal corresponds to one's hand turning the tap to open and close it. The output signal is the corresponding variation in the flow from the tap — but this can only occur if the tap is connected to its power supply — the water main. Without that power supply, there is no output, no matter how great the input.

Furthermore, the pressure of the output from the tap can be no greater than the supply pressure — in fact it is always a bit less. In the same way, the output amplitude of an electronic amplifier can never be greater than its power supply voltage. This is illustrated in Fig. 3.11. Suppose the operational amplifier shown in Fig. 3.10 has a gain of 1000, and we apply an a.c. voltage with an amplitude of 10 mV to its input. The output voltage will have an amplitude of $1000 \times 0.01 = 10$ V, and, this being well within the power supply voltage, a faithful copy of the input waveform will appear in amplified form at the output, as shown in Fig. 3.11(a).

Fig. 3.11 An amplifier's output waveform when the amplitude is (a) less than and (b) greater than the power supply voltage

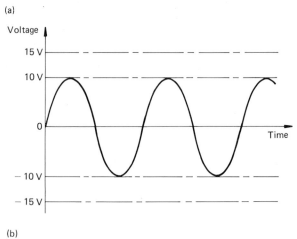

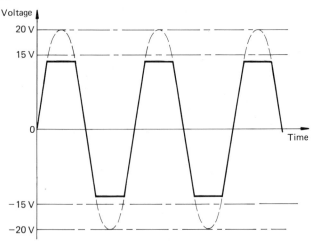

If, however, we increase the input amplitude to 20 mV, so that the output amplitude should be $1000 \times 0.02 = 20$ V, the output will be limited or *clipped* to about 14 V (the power supply voltage less the voltage drop inside the amplifier). As Fig. 3.11(b) shows, the effect is to cut off the top and bottom of the waveform, distorting it into something like a square wave. If the input is a sine wave of audiofrequency, and the output is taken to a power amplifier and loudspeaker, the effect of the clipping can be heard as a 'rattle' superimposed on the pure note of the sine wave.

STABILISING AN AMPLIFIER

An electronic amplifier is a form of control system — in which the output is controlled by the input. In fact, methods used to determine the stability of control systems were originally devised, some fifty years ago, to determine the stability of telecommunications amplifiers.

A voltage amplifier, such as the operational amplifier described on p. 57, may have a gain of 100 000 or more. If we were to use such a high-gain amplifier simply as shown in Fig. 3.9, we would, in effect, be using an open-loop control system. Random variations in the flow of electrons within the amplifier would cause the gain to vary from one instant to the next, so that a sine wave input, being amplified in random fashion, could become hopelessly distorted.

To stabilise such an amplifier, we must turn it into a closed-loop control system with negative feedback. (These terms are explained in Chapter 7.) The general principle is shown in Fig. 3.12, in which A is the amplifier, and b is some arrangement of resistances which feeds back a small proportion of the output voltage to be subtracted from the input voltage. This must inevitably cut down the gain available from the amplifier, but it does keep it constant. For example, if a momentary fall in amplifier gain causes the output voltage v_o to start to fall, this diminishes v_f, with the result that the actual input to the amplifier, $v_i - v_f$, is increased to compensate.

Fig. 3.12 Amplifier, stabilised by a closed-loop negative feedback system

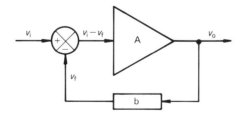

Fig. 3.13 shows how this principle is put into effect in the case of an operational amplifier. R_f is the feedback resistor corresponding to the block marked b in Fig. 3.12. To obtain *negative* feedback it is connected to the *inverting* input of the amplifier.

Now it can be shown mathematically that if the gain of the amplifier itself is 'very large', the gain of the circuit shown in Fig. 3.13 is given by the ratio R_f/R_1.

Fig. 3.13 Stabilising and
setting the gain of an
operational amplifier

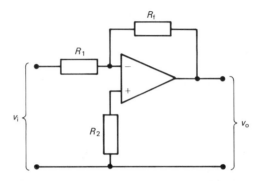

Thus, no matter how the gain of the amplifier changes (within reason), the output voltage v_o will always be $(R_f/R_1) \times v_i$. This is a great improvement, because we can set the gain of an amplifier by setting the value of R_f, and that gain will remain constant, since resistors are stable, passive components.

Resistor R_2 should have the same resistance as R_1 and R_f in parallel; i.e. for practical purposes $R_2 = R_1$ if R_f is very much greater than R_1.

R_1 and R_2 are usually in the range 1–10 kΩ. Suppose we make them 10 kΩ each. Then to obtain a stabilised gain of 20, for example, R_f should be $20 \times R_1 = 200$ kΩ.

SOME APPLICATIONS OF OPERATIONAL AMPLIFIERS

Summing

Any number of signals can be added together in various proportions by bringing each signal through its own particular resistor to the inverting input of an operational amplifier, as shown in Fig. 3.14. The output voltage is then the sum, for all the input signal voltages, of

$$\text{Input signal voltage} \times \frac{\text{Feedback resistance}}{\text{Input resistance}}$$

That is, referring to Fig. 3.14:

$$v_0 = -\left(\frac{R_f}{R_1}v_1 + \frac{R_f}{R_2}v_2 + \ldots + \frac{R_f}{R_n}v_n\right)$$

Sign Reversal

When a signal is applied to the inverting input of an operational amplifier, its sign is reversed — that is, a positive voltage at input gives a negative output voltage; similarly a negative input gives a positive output. To restore the original polarity of the signal, we can pass it through another operational amplifier with a voltage gain of 1. This would be done by making R_f equal to R_1 (see Fig. 3.13) for the second operational amplifier.

Fig. 3.14 A summing amplifier, for adding signals v_1, v_2, ... and v_n

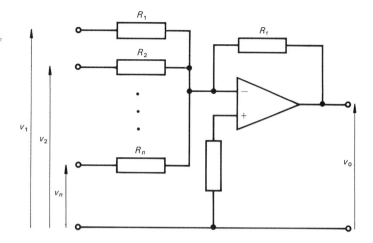

Signal Comparison

We compare two signals by subtracting one signal from the other. This can be done by applying one signal to the inverting input and the other to the non-inverting input of an operational amplifier, as shown in Fig. 3.15. If $R_1 = R_3$ and $R_2 = R_4$, the output voltage is

$$v_0 = \frac{R_2}{R_1}(v_B - v_A)$$

A typical application would be to amplify the output of a bridge circuit such as the strain-gauge bridge in Fig. 3.20. The galvanometer would be replaced by the operational amplifier circuit of Fig. 3.15, with input A connected to point P, and input B connected to point Q. The amplifier will then amplify the voltage difference between P and Q while ignoring any voltage change common to both P and Q, such as would occur if the power supply voltage altered.

Fig. 3.15 A difference amplifier for amplifying the difference between v_A, and v_B

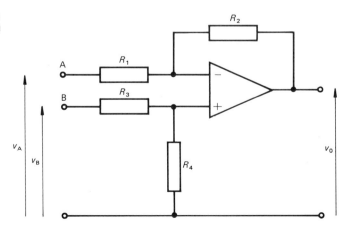

FREQUENCY RESPONSE CURVES

For a simple mechanical device, such as a spring or a gear train, the gain is a constant which is fixed by the designer, being determined by the relationship:

$$\text{Gain} = \frac{\text{Change of output value}}{\text{Corresponding change of input value}}$$

However, for electronic amplifiers, and for many other items of signal-conditioning or transducer equipment, the gain varies with the frequency of the alternating input applied to the component. In these cases it can only be specified by drawing a graph of gain (vertically) against frequency (horizontally), as shown in Fig. 3.16.

Fig. 3.16   A typical frequency response curve

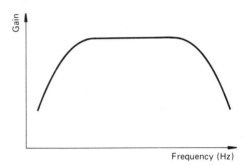

GAIN IN DECIBELS

In practice, gain is not plotted as a simple ratio, but as a number of decibels (abbreviation: dB), calculated from the following formula:

$$\text{Gain (dB)} = 20 \log_{10}\left(\frac{V_{\text{out}}}{V_{\text{in}}}\right)$$

This can be thought of as

$$10 \times \log_{10}\left(\frac{V_{\text{out}}}{V_{\text{in}}}\right)^2.$$

In other words, we are taking gain as the ratio of output to input voltage, squaring it (because electrical power is proportional to voltage squared), taking the log-to-the-base-ten of the result (which would give us gain in *bels*), and then multiplying by ten to get decibels.

There are two reasons for taking the log of the gain ratio:

(i) If we want the overall gain of a series of amplifiers, we have to multiply their individual gains together. If the gains are defined by graphs we would have to plot a graph of overall gain by multiplying together the corresponding ordinates of the individual graphs — which would be quite a job! If, however, we are plotting the *logarithms* of gain, we only have to *add* the ordinates, which can easily be done on the same graph sheet.

(ii) Since decibels are a logarithmic quantity and since we always plot frequency on a logarithmic scale as well (*i.e.* frequency increases ten-fold at constant intervals), we are plotting log against log, and this has the effect of converting most frequency response curves into straight lines, or at least into shapes which can be approximated to straight lines.

So to get the overall gain of a series of amplifiers, we add their gains in decibels. This is what makes the decibel scale so useful, but it takes a bit of getting used to. The following table of numerical values of gain and their corresponding decibel values may help to make it clearer.

dB gain	Numerical gain $\left(= \dfrac{V_{out}}{V_{in}} \right)$
100	100 000
80	10 000
60	1 000
40	100
20	10
6	2
3	1.41 (i.e. $\sqrt{2}$)
0	1
-3	0.707 $\left(\text{i.e. } \dfrac{1}{\sqrt{2}} \right)$
-6	0.5 $\left(\text{i.e. } \dfrac{1}{2} \right)$
-20	0.1 $\left(\text{i.e. } \dfrac{1}{10} \right)$

BANDWIDTH

Since the electrical power developed across a resistance is proprotional to the square of the voltage drop, a fall of 3 dB in amplifier gain means that the output power has been multiplied by $(1/\sqrt{2})^2$; that is, it has been halved. The *bandwidth* of an amplifier is therefore taken as the range of frequencies lying between the 3 dB points — that is, between the two frequencies at which the gain is 3 dB less than the maximum (see Fig. 3.17). These two frequencies are sometimes called the *cut-off frequencies*.

Fig. 3.17 How the
bandwidth of an
amplifier is measured

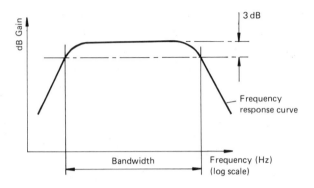

SELF-TEST QUESTION 6 (Solution on p. 234)

Three amplifiers are arranged in series (i.e. the output of the first
becomes the input to the second, and the output of the second
becomes the input to the third). At a particular frequency the
amplifiers have voltage gains of 12, 25 and 4, respectively.
Determine the overall gain by multiplying together their voltage
gains, and check your answer by adding up their dB values of gain.

AN AMPLIFIER PROJECT

For a proper understanding of electronic amplifiers in general and
of the material on pp. 55–63 in particular, students should construct and test their own amplifiers. This could be an assignment
under Section B (Measurement Systems) of the TEC Unit, and it
would form a useful introduction to the practical skills required in
instrumentation construction and testing. Full details of a suitable
project are given in Chapter 6.

THE WHEATSTONE BRIDGE

Developed by Sir Charles Wheatstone in the nineteenth century,
the Wheatstone bridge is an electrical circuit for measuring
resistance accurately. Fig. 3.18 is the circuit diagram. It consists
of four resistances, a galvanometer and a d.c. power supply.

Fig. 3.18 The
Wheatstone bridge

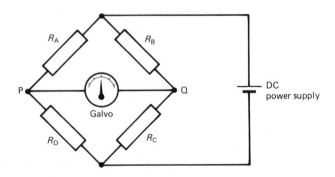

R_A is the resistance to be measured, R_D is a fixed (i.e. constant) resistance and the ratio R_B/R_C is adjustable, either by making R_B or R_C a variable resistance, or by making $R_B + R_C$ one continuous fixed resistance with a variable tapping point for the galvanometer connection.

The galvanometer is a very sensitive moving-coil meter with centre-zero. (This means that the scale is graduated in a number of equal divisions with 0 (zero) in the middle of the scale, and the pointer normally points to zero when the instrument is not in use.)

To use the bridge circuit to measure R_A, we must first *balance the bridge*. This is done by adjusting the ratio R_B/R_C until the galvano-meter indicates zero. When the galvanometer indicates zero, no current is passing through it, and this indicates that there is no voltage difference between its terminals (i.e. that the voltages at points P and Q in Fig. 3.18 are equal).

Now R_A and R_D are carrying the same current, so

$$\text{Voltage at P} = \frac{R_D}{R_A + R_D} \times \text{Power supply voltage}$$

Also R_B and R_C are carrying the same current, so

$$\text{Voltage at Q} = \frac{R_C}{R_B + R_C} \times \text{Power supply voltage}$$

Therefore, when the bridge is balanced (galvo reading zero):

$$\frac{R_C}{R_B + R_C} = \frac{R_D}{R_A + R_D}$$

$$\therefore \quad R_C(R_A + R_D) = R_D(R_B + R_C)$$

$$\therefore \quad R_A R_C + R_C R_D = R_B R_D + R_C R_D$$

$$\therefore \quad R_A R_C = R_B R_D$$

$$\therefore \quad \frac{R_A}{R_D} = \frac{R_B}{R_C}$$

Thus

$$R_A = R_D \times \frac{R_B}{R_C}$$

This result is independent of the power supply voltage.

EXAMPLE   Find R_A if $R_C = 180\,\Omega$, $R_D = 390\,\Omega$, and R_B had to be adjusted to $227.3\,\Omega$ to balance the bridge.

SOLUTION

$$R_A = R_D \times \frac{R_B}{R_C}$$

$$= 390 \times \frac{227.3}{180} \, \Omega$$

$$= 492 \, \Omega$$

THE WHEATSTONE BRIDGE AS A LABORATORY EXPERIMENT

To 'get the feel of' the Wheatstone bridge circuit, it is useful to carry out a few resistance measurements using the laboratory form of the Wheatstone bridge. This is shown in Fig. 3.19:

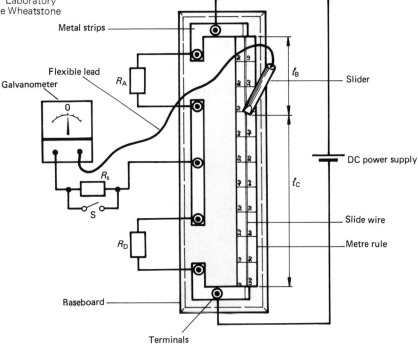

Fig. 3.19 Laboratory form of the Wheatstone bridge

The resistances are connected together by metal strips fixed to a baseboard. R_A, the unknown, and R_D, the known resistance, are connected across gaps in the strips. $R_B + R_C$ is a single length of resistance wire (the slide wire) stretched along a metre rule. The metre rule is graduated from 0 to 100 from both ends, so that the distance of any point from either end can be read off directly in centimetres.

To protect the galvanometer from being burnt out by the accidental application of an over-large voltage, it is connected to the bridge through a series resistor, R_S, of a few thousand ohms. R_S can be 'shorted out' by closing switch S, to give the full sensitivity of the galvanometer, when the approximate balance point on the slide wire has been found.

To measure R_A, the slider is slid along the slide wire, with switch S open, until the galvanometer reads zero. Switch S is then closed and the slider position readjusted, if necessary, to return the galvanometer reading to zero. The bridge is now balanced. Then

$$\frac{R_A}{R_D} = \frac{R_B}{R_C}$$

$$= \frac{l_B}{l_C}$$

since the resistance of the slide wire is proportional to its length. Therefore

$$R_A = R_D \times \frac{l_B}{l_C}$$

(To avoid getting the length ratio upside down, remember that l_B is alongside R_A and l_C alongside R_D.)

EXAMPLE   If R_D is 46.2 Ω, and the bridge balances with the slider in the position shown in Fig. 3.19, estimate the resistance of R_A.

SOLUTION   From Fig. 3.19, l_B and l_C are estimated as 32 cm and 68 cm respectively. Therefore

$$R_A = 46.2 \times \frac{32}{68} \, \Omega$$

$$= 21.7 \, \Omega$$

It is instructive to measure a few resistances in this way, and also to vary the voltage supplied to the bridge, if a variable voltage d.c. power supply is available. You should find that altering the supply voltage has no effect on the output value of the measurement system (R_A) but alters the resolution with which it can be obtained. (Why?)

Of course, if we are to measure R_A accurately, we must have an accurate value of R_D by some other means, before we start.

THE WHEATSTONE BRIDGE APPLIED TO STRAIN GAUGES

Actually, in instrumentation, we are not interested in using the Wheatstone bridge for 'coarse' measurements of resistance such as we have considered in the last two examples. We use it, instead, as the signal conditioner for resistance transducers such as the resistance thermometer or the strain gauge. In transducers such as these, the change of resistance which constitutes the output of the transducer is very small compared with the resistance of the transducer itself. By balancing it against an identical fixed resistance in a Wheatstone bridge circuit, we 'cancel out' the original resistance of the transducer, leaving only the small change of resistance. Similar circuits with an a.c. power supply are used to measure the small changes of inductance or capacitance produced by inductive or capacitive transducers. All such circuits are called *bridge circuits*. You have already met an application of this principle in the linear variable differential transformer of pp. 41-3.

When the Wheatstone bridge circuit is used as the signal conditioner for a strain gauge, the circuit of Fig. 3.18 still applies, but R_A and R_D are now strain gauges. R_A is the *active gauge*, bonded to the material whose strain is to be measured. R_D is the *dummy gauge*, an identical strain gauge bonded to a loose unstrained piece of the same material, which is placed as close as possible to R_A so that it shares the same temperature, and therefore can cancel out the effects of any temperature change. The temperature change may be due to:

(a) the heating effect of the current through the strain gauge, or

(b) a change in air temperature — for example, a cold draught blowing over the strain gauge when doors are left open. The effects of such temperature change are:

   (i) a change of electrical resistance due to the strain gauge's behaving like a resistance thermometer;

   (ii) expansion or contraction of the material to which the gauge is bonded, causing a temperature-induced strain, which may be mistaken for stress-induced strain.

Temperature-compensated gauges are available to reduce or eliminate the above errors. There are normally two types, one type correcting for the linear expansion of steel, the other for aluminium. However, it is advisable still to use a dummy gauge, instead of a fixed resistor for R_D, wherever possible, to make sure that all temperature effects are eliminated.

Fig. 3.20 shows the active and dummy strain gauges incorporated into the basic bridge circuit of Fig. 3.18. R_B and R_C are approximately equal resistances; the bridge is balanced by adjusting R_B so that the galvanometer reads zero.

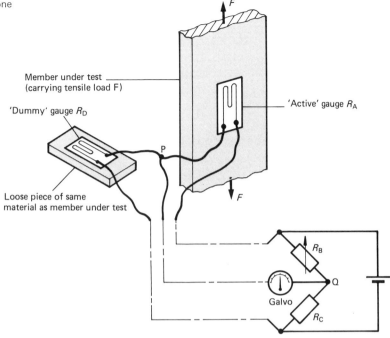

Fig. 3.20 Wheatstone bridge as used for strain-gauging

Let δR_A denote a change in the resistance of the active gauge, and δR_B be the change in R_B after the bridge has been rebalanced. Then,

$$R_A = R_D \times \frac{R_B}{R_C} \qquad [1]$$

and

$$R_A + \delta R_A = R_D \times \frac{(R_B + \delta R_B)}{R_C}$$

$$\therefore \qquad R_A + \delta R_A = R_D \times \frac{R_B}{R_C} + R_D \times \frac{\delta R_B}{R_C} \qquad [2]$$

Subtracting equation [1] from equation [2] gives

$$\delta R_A = \frac{R_D}{R_C} \times \delta R_B$$

If R_D and R_C are approximately equal, this simplifies to

$$\delta R_A = \delta R_B$$

EXAMPLE In a test using the circuit shown in Fig. 3.20 the active and dummy gauges were $120\,\Omega$ strain gauges with gauge factor 2.1, and R_C was a $600\,\Omega$ resistor. R_B was $601.2\,\Omega$ when the bridge originally balanced; a steady load was then applied to the member under test, and the bridge was rebalanced — this required the adjustment of R_B to $603.8\,\Omega$. Calculate the strain in the member under test.

SOLUTION   The change in resistance of the active gauge is given by

$$\delta R_A = \frac{R_D}{R_C} \times \delta R_B$$

$$= \frac{120}{600} \times (603.8 - 601.2)$$

$$= 0.52 \, \Omega$$

Then, using the equation given on p. 32.

$$\frac{\delta R_A}{R_A} = k\epsilon$$

$$\therefore \quad \frac{0.52}{120} = 2.1\epsilon$$

$$\therefore \quad \epsilon = \frac{0.52}{120 \times 2.1}$$

$$= 0.002\,06 \quad \text{or} \quad 2060 \text{ microstrain}$$

Actually we can obtain the same result more simply by pretending that we are using strain gauges of the same resistance as R_B, and that the change in R_B was also the change in resistance of the active gauge.

Then $$\frac{\delta R}{R} = k\epsilon$$

becomes $$\frac{(603.8 - 601.2)}{601.2} = 2.1\epsilon$$

$$\therefore \quad \epsilon = \frac{2.6}{601.2 \times 2.1}$$

$$= 0.002\,06$$

SELF-TEST QUESTION 7 (Solution on p. 235)

In a test to determine the modulus of elasticity of an aluminium alloy, a test specimen with a circular cross-section 13.9 mm in diameter was strain-gauged as in Fig. 3.20 and loaded in tension. The resistances of all four arms of the bridge were nominally 800 Ω each, and the strain gauges had a gauge factor of 2.05. The following results were obtained:

Load (kN)	0	2	4	6	8	10	12	14	16	18	20
Resistance R_B at balance (Ω)	800.1	800.3	800.6	801.0	801.2	801.5	801.9	802.1	802.4	802.7	803.0

Plot a graph of the results, and from the graph estimate the modulus of elasticity of the material.

THE DIRECT-READING BRIDGE

When the Wheatstone bridge is balanced so that the galvanometer reads zero, it is said to be in the *null condition*, and this method of using the bridge is called the *null method*. It has the advantage that the result is unaffected by any possible variation in the voltage of the power supply, but it is a slow, cumbersome method which cannot give a continuous reading of a varying strain-gauge resistance, and hence it would be useless for building into an automatic control system.

Nowadays, stabilised d.c. power supplies are available, which are powered by a.c. mains and which give a d.c. output of virtually constant voltage. It is better to use one of these power supplies, and leave the bridge unbalanced, except for balancing it by adjusting R_B at zero strain, at the start of a test. We can obtain a calibration of galvo reading against change in R_A by switching-in parallel calibration resistors to artificially change the value of R_A.

This circuit is shown in Fig. 3.21. For an automatic control system, the galvanometer could be replaced by an operational amplifier.

Fig. 3.21 Direct-reading bridge

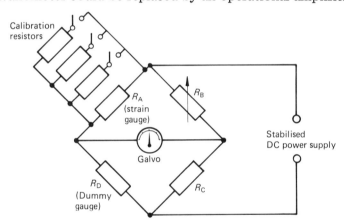

STRAIN-GAUGE SYSTEMS

To fully strain-gauge a complex structure may require many pairs of strain gauges. Obviously it would be uneconomic to build each pair into a separate bridge circuit with its own galvanometer. Also the response time of a galvanometer is far too slow to follow strain oscillations caused by vibration. So we need a single display unit, with a more rapid response — an oscilloscope would be ideal here. But this will require an amplifier to amplify the unbalance signal from the bridge.

We also want to be able to switch this unit (comprising $R_B + R_C$, power supply, amplifier and display device) on to each pair of gauges R_A and R_D in turn. We can put all the pairs of gauges in parallel; this puts an additional load on the power supply but otherwise does not affect the circuit — the only pair of gauges forming part of the bridge circuit is the pair whose centre junction is switched into the amplifier at that time.

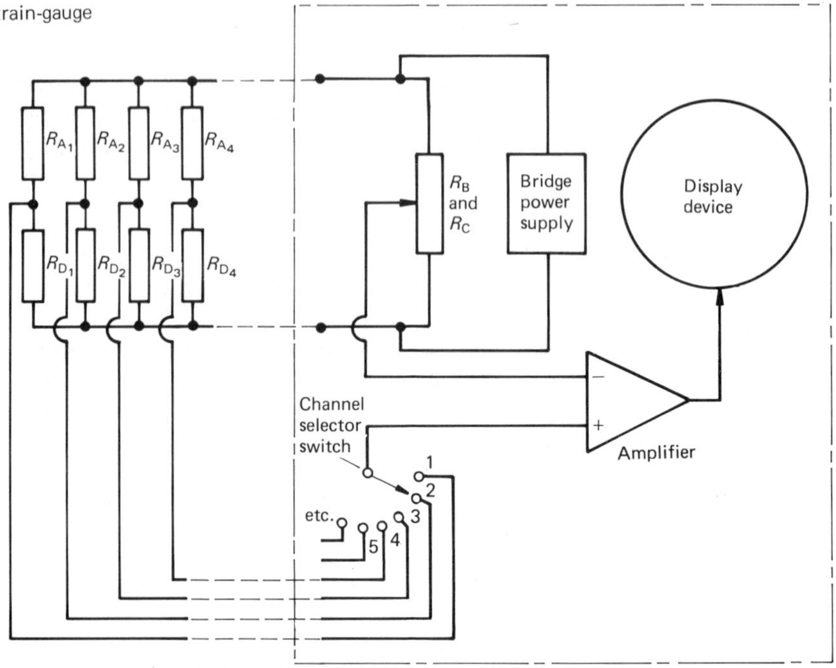

Fig. 3.22 Strain-gauge
oscilloscope

The result is the strain-gauge oscilloscope, shown diagrammatically in Fig. 3.22.

In Fig. 3.22, the channel selector is shown switched to channel 2 — i.e., connecting the centre junction of gauges R_{A_2} and R_{D_2} to the amplifier input. The other gauges have no effect on the signal from this pair. We can switch over to any other pair at any instant; meanwhile they all have power supply current passing through them, keeping them warm, and so eliminating any warming-up temperature errors when we switch channels.

The amplifier amplifies the difference in voltage between its positive and negative inputs, to operate the display device.

In a commercial strain-gauge oscilloscope we shall probably find that the power supply is low-voltage a.c. instead of d.c., and that resistance $R_B + R_C$ is replaced by an inductance, with the balance point selected by tappings off the inductance. And the tapping points will probably be already graduated in microstrain. No matter — these are refinements which do not affect the basic principle illustrated in Fig. 3.22.

GETTING RID OF THE DUMMY

To increase the sensitivity of a strain-gauge system, to do away with the untidiness of loose pieces of metal, and to get the dummy gauge situated even closer to the active gauge, we usually try to turn the dummy into another active gauge, by sticking it on to the specimen in such a way that it will add to the signal from the active gauge. Various ways of doing this are shown below.

Tension. The dummy gauge is placed where it will be acted upon by the compressive strain caused by the Poisson's ratio effect, as shown in Fig. 3.23(a). This multiplies the strain sensitivity of the bridge by a factor of $(1 + \nu)$, where ν is Poisson's ratio for the material to which the gauges are stuck.

Bending. The active and the dummy gauge are placed on opposite sides of the neutral axis, as shown in Fig. 3.23(b). For a cross-section symmetrical about the neutral axis, this doubles the strain sensitivity and makes the arrangement insensitive to any end load which may be present, as the signals due to end load from the two gauges cancel out.

Shear (including shear due to torsion). The active and the dummy gauges are placed at 45° to the direction of the shear and at 90° to each other, as shown in Fig. 3.23(c), so that one is acted on by the tensile, and the other by the compressive strain set up by the shear.

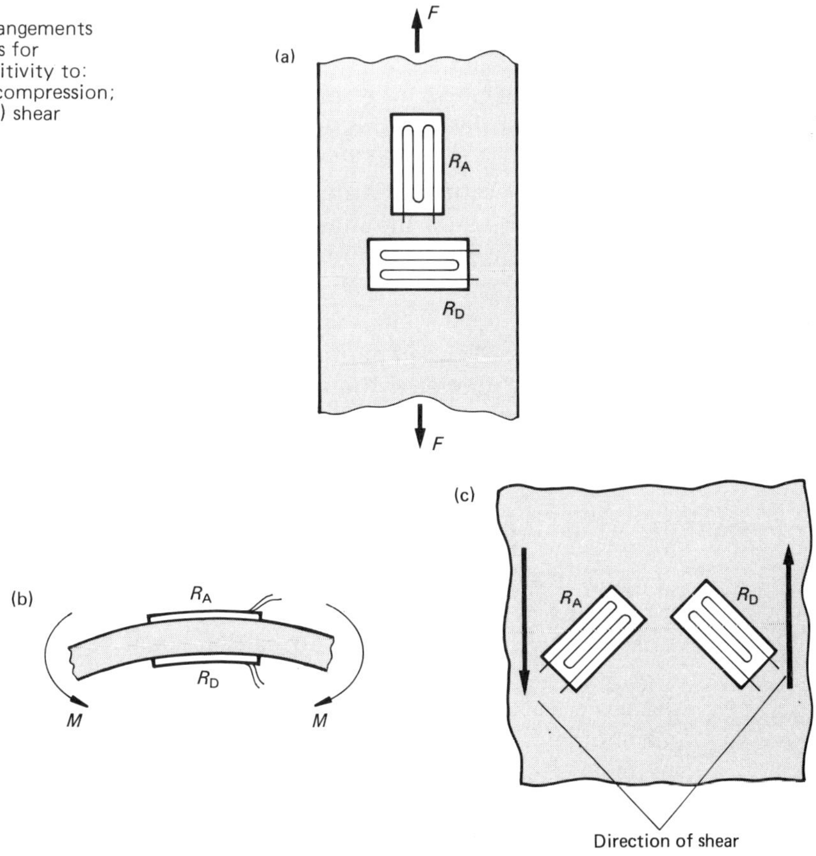

Fig. 3.23 Arrangements of strain gauges for maximum sensitivity to: (a) tension or compression; (b) bending; (c) shear

LOAD MEASUREMENT

When strain gauges are used for *load* measurement, they are calibrated, if this is at all possible, by applying known loads to the strain-gauged part and noting the corresponding bridge output readings. The resulting calibration curve can then be used to determine unknown loads by reading them off against the corresponding bridge output values. Thus the type of calculation illustrated by Self-Test Question 1 on p. 32 is eliminated whenever possible, because it assumes the stress to be uniformly distributed over the cross-section. While this may be true for the simple tie-bar in that example, it would be untrue for more complicated structures.

THE WHEATSTONE BRIDGE APPLIED TO TEMPERATURE MEASUREMENT: THE RESISTANCE THERMOMETER

The property which is undesirable in a strain gauge — sensitivity to temperature change — can be made use of for accurate temperature measurement. Most metals increase in resistance as their temperature increases. Between two adjacent points on the International Practical Temperature Scale (which is a series of exactly reproducible temperatures fixed by the freezing or boiling points of various substances) the resistance change per degree of temperature change (known as the *temperature coefficient of resistance* of a metal) can be taken as constant. Thus an electrical resistor can act as a temperature transducer, the temperature difference from the nearest point on the International Practical Temperature Scale being proportional to the resistance difference from the resistance at that temperature.

Metals commonly used as temperature sensors are copper, nickel, platinum and tungsten, the last three being particularly useful for high temperature measurement because of their high melting points.

As in the case of strain gauges, the resistance change is small, so a Wheatstone bridge circuit must be used to measure it. This time there is no 'dummy' sensor, but the resistance change in the connecting wires caused by the temperature gradient along them must be cancelled out, by ensuring that equal lengths of connecting wire with equal temperature gradients occur in series with both the temperature sensor and the resistance on the other side of the galvanometer connection. So the circuit for a resistance thermometer is either as shown in Fig. 3.24 or as in Fig. 3.25. Again, for automatic control, the galvanometer could be replaced by an operational amplifier.

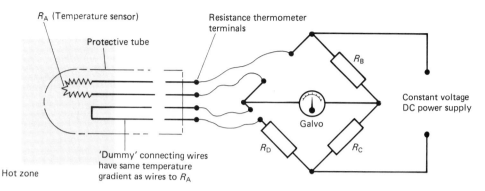

Fig. 3.24 Resistance thermometer circuit — four-wire system

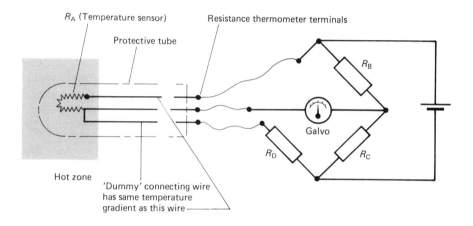

Fig. 3.25 Resistance thermometer circuit — three-wire system

THE POTENTIOMETER

In Fig. 3.26 the symbol inside the dotted lines is the circuit diagram representation of a potentiometer. The *wiper* CX may be moved to any point along the resistor AB.

Provided that no current is taken from C, the output voltage V_o is given by

$$V_o = \frac{\text{Resistance from A to X}}{\text{Resistance from A to B}} \times V_s$$

where V_s is the voltage applied to AB. If the resistor AB is *linear* (i.e. if AB has a constant ratio of resistance to length at every point along its length), we can further say that

$$V_o = \frac{\text{Length AX}}{\text{Length AB}} \times V_s$$

Fig. 3.26 Potentiometer

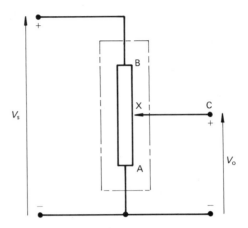

If some current is drawn from C (for example, by measuring V_o with a voltmeter), this equation is no longer quite true. This is because portion XA is carrying only the current drawn from A, whereas portion BX is carrying both the current drawn from A and the current drawn from C. (Current is assumed to flow from + to −.) So by Ohm's law ($V = IR$), the extra current drawn from C causes the voltage drop from B to X to be greater, and the voltage drop from X to A to be less than they otherwise would be.

In practice, the current taken from C is usually small enough for us to be able to ignore the non-linearity it produces in the relationship between length AX and output voltage V_o.

THE POTENTIOMETER AS A TRANSDUCER

The form of potentiometer used in the electronics industry is shown in Fig. 3.27. This component is also used as a variable resistor when required, by connecting to the wiper and to only

Fig. 3.27 Potentiometer. The wiper is always the middle of the three connections

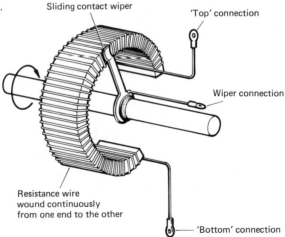

Sliding contact wiper

'Top' connection

Wiper connection

Resistance wire wound continuously from one end to the other

'Bottom' connection

one end of the resistance. In such a case, it is usual to connect the other end of the resistance back to the wiper, so that the unused portion of the resistance is short-circuited by the wiper, instead of being left 'floating'.

A wire-wound potentiometer is shown — this would be used where a hard-wearing resistance element is needed, to withstand constant rubbing of the wiper. It has the disadvantage that the voltage picked off by the wiper changes in discrete (i.e. separate) steps, as shown in Fig. 3.28, each step corresponding to the wiper's contacting the next turn of resistance wire.

Fig. 3.28 Voltage/displacement response of a wirewound potentiometer

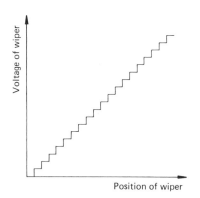

The alternative is a *carbon track* potentiometer, in which the resistance element is moulded from a carbon composition. These are high resistance potentiometers (typically $5\,\text{k}\Omega$ to $2\,\text{M}\Omega$). They are cheap, compact, and the voltage picked off by the wiper varies smoothly and continuously without any steps, but the carbon track sometimes acquires a tarnish as it ages, so that in time it becomes 'noisy' in operation. A modern improvement is the *cermet track* potentiometer in which the resistance element is a cermet (sintered powdered metal and ceramic material). These are available in a very wide range of resistances, and have very good electrical and thermal stability together with good linearity and temperature coefficient.

When a potentiometer is used as a position transducer, it has a d.c. voltage from a power supply applied to the ends of the resistance track, and it converts 'position of wiper' into a proportional voltage between the wiper and one end of the track.

THE CHARGE AMPLIFIER

The charge amplifier has already been mentioned in Chapter 2 (p. 39) as being the proper amplifier to use with a piezoelectric transducer. Now, in this chapter on signal conditioning equipment, we should consider why this is so, and what makes a charge amplifier different from any other kind of amplifier.

As stated in Chapter 2, the input to a piezoelectric transducer is force, and its output is charge. This *charge* is in the form of an increased population of electrons on one of the electrodes of the crystal, and a correspondingly decreased population on the other electrode. Its effect is a voltage between the two electrodes, determined by the equation of a capacitor:

$$v = \frac{q}{C}$$

where q is the charge in coulombs and C is the capacitance in farads.

The electrical connections to the two electrodes of a piezoelectric transducer are always made using *coaxial cable*, because in this type of cable the outer conductor screens the inner conductor from picking up electrical 'noise'. The outer conductor is a sleeve woven from copper wire; the inner conductor is a wire which runs down the centre of the sleeve and is insulated from it by a high-grade insulating material such as polythene. (TV aerial cable is an example of a coaxial cable.)

If we plug the other end of the coaxial cable into the input socket of an ordinary amplifier, the capacitance between the conductors of the cable adds to the capacitance of the crystal, so, as we can see from the equation above, the output voltage of the crystal will be reduced. Also the input resistance of an ordinary amplifier is usually so low that the charge on the crystal will leak away very quickly indeed.

A charge amplifier gets round these difficulties by having a *very* high input resistance, and by using a type of circuit known as a Miller integrator, which makes the output voltage independent of the capacitance of the crystal and cable.

Fig. 3.29 Piezoelectric transducer and charge amplifier

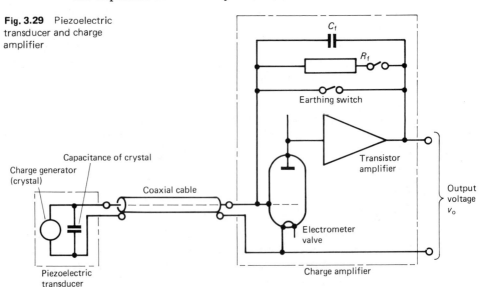

Fig. 3.29 shows the principle of the charge amplifier.

The output of the piezoelectric crystal is applied to the grid of an electrometer valve. This is like a radio valve, but so constructed and mounted that its input resistance is of the order of $10^{14}\,\Omega$. It boosts the voltage of the signal from the crystal but more important, it supplies the current to maintain that voltage when it is fed into the transistor amplifier which follows the valve.

The effect of the feedback capacitor, C_f, between the output and the input, is to automatically adjust the gain of the amplifier to make it proportional to whatever capacitance of crystal and cable we have on the input side. So we can use any length of coaxial cable we like (within reason) for connecting the crystal to the charge amplifier, because the output voltage of the crystal-and-charge-amplifier combination is defined by:

$$v_0 = \frac{q}{C_f}$$

where q is the charge produced by the force on the crystal and C_f is the feedback capacitance. C_f can be made as large or as small as we like.

Two more features of Fig. 3.29 are worth noting: the earthing switch, and the bypass resistor, R_f. The earthing switch (usually a push-button) is closed for an instant when we want to 'zero' the system. It gets rid of any static charge which may have accumulated on the crystal or on the amplifier input. The bypass resistor R_f is switched in parallel with C_f when we want to measure dynamic (alternating) rather than static forces on the crystal. It provides a leakage path for the charge on the crystal. This reduces the time constant of the system but allows it to follow rapidly changing forces more accurately.

The details of R_f and C_f are taken care of by the manufacturer of the charge amplifier, who usually provides a range switch, which switches into the circuit paired values of R_f and C_f to suit the signal strength expected from the transducer.

Anyway, the essential thing we must remember about a charge amplifier is that its input is charge and its output is voltage.

EXAMPLE   A piezoelectric force transducer has a charge sensitivity of $20\,\text{pC/N}$. It is to be connected to a charge amplifier, and the combination of transducer and amplifier is to have an overall gain of $50\,\text{mV/N}$.

(a) What gain should the amplifier have?

(b) What approximate value of feedback capacitor is needed to give this gain? ($1\,\text{pC}$ is 10^{-12} coulomb.)

SOLUTION   (a) Construct a block diagram as in Fig. 3.30.

Combining the blocks gives

$$20 \times x = 50$$

$$\therefore \qquad x = \frac{50 \, \text{mV/N}}{20 \, \text{pC/N}}$$

$$= 2.5 \, \frac{\text{mV}}{\text{pC}}$$

Therefore the gain of the amplifier should be 2.5 millivolts per picocoulomb.

Fig. 3.30 See example

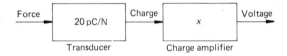

Force → | 20 pC/N | → Charge → | x | → Voltage

Transducer   Charge amplifier

(b) $$v_o = \frac{q}{C_f}$$

Putting the data into basic SI units gives

$$2.5 \, \frac{\text{mV}}{\text{pC}} = \frac{2.5 \times 10^{-3}}{10^{-12}} \, \frac{\text{V}}{\text{C}} = 2.5 \times 10^{9} \, \frac{\text{V}}{\text{C}}$$

From the block diagram we can see that

$$\text{Gain of charge amplifier} = \frac{\text{Output}}{\text{Input}}$$

$$= \frac{\text{Output voltage}}{\text{Charge}}$$

Then $$2.5 \times 10^{9} = \frac{v_o}{q}$$

$$= \frac{1}{C_f} \quad \text{from the equation above.}$$

$$\therefore \qquad C_f = \frac{1}{2.5 \times 10^{9}} \, \text{F}$$

$$= 0.4 \times 10^{-9} \, \text{F}$$

$$= 400 \, \text{pF}$$

MODULATION

If we listen to radio or watch television, the output device, a loudspeaker or a cathode-ray tube, is operated by an analogue signal (the word *analogue* has already been explained on p. 21). The receiver is usually many miles from the radio or TV transmitter and is capable of being *tuned* so that it picks up and amplifies the required transmission while rejecting all the other transmissions which it is capable of receiving.

This can only be done if the transmission is in the form of a very high frequency a.c. voltage in the aerial of the transmitter, and the tuned circuits of the receiver are tuned so that they resonate electrically at that frequency.

Now the analogue signal needed to operate our loudspeaker or TV tube has no constant frequency, but consists of many continually varying frequencies, all of which are very much lower frequencies than that required for radio or TV transmission. The information in our analogue signal therefore has to be *modulated* on to a *carrier wave*, which is at the radio or TV transmission frequency. This carrier wave is radiated from the TV or radio transmitter, and picked up by the receiver. When the tiny carrier wave voltage picked up by the receiver has been amplified up to a reasonable strength, it is *demodulated*, to get rid of the carrier wave and leave only the analogue signal which was superimposed on it.

Fig. 3.31 Carrier wave

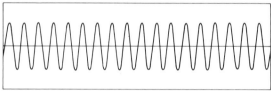

Fig. 3.32 Analogue signal

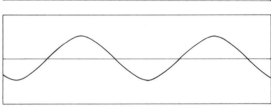

Fig. 3.33 Amplitude modulation: amplitude of carrier wave modulated by signal voltage

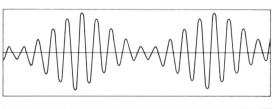

Fig. 3.34 Frequency modulation: frequency of carrier wave modulated by signal voltage

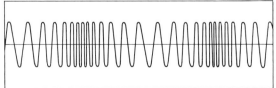

The two main methods of modulation are *amplitude modulation* (AM) and *frequency modulation* (FM). These two methods are illustrated in Figs. 3.31 to 3.34, which represent graphs on which

voltage is measured along the vertical axis and time along the horizontal axis.

Fig. 3.31 shows a typical carrier wave, and Fig. 3.32 an analogue signal which is to be modulated on to it.

Fig. 3.33 shows how the analogue signal is superimposed on the carrier wave by amplitude modulation, making its amplitude (i.e. height) vary as the signal voltage varies.

Fig. 3.34 shows how frequency modulation works. In this case the frequency is increased as the signal voltage rises and decreased as it falls.

Amplitude modulation is used for all long, medium and short wave broadcasting. Frequency modulation requires a very high frequency (VHF) carrier wave, as the frequency variation is actually very small in relation to the basic carrier wave frequency. The advantage of this type of modulation is that it gives high quality, low interference reception, since the interference and distortion effects picked up by a receiver are in the form of changes of amplitude, to which an FM receiver is insensitive.

Either type of modulation may be used in *telemetry*, in which a continuous measurement of some quantity such as temperature is transmitted by radio from a moving object such as a rocket, to a receiver on the ground. A modulated carrier wave may also be used where the torque or stress in a rotating shaft has to be measured. Another example may be found in Fig. 4.10 where the engine indicator system uses a frequency-modulated a.c. carrier for the 'volume' signal.

ANALOGUE-TO-DIGITAL AND DIGITAL-TO-ANALOGUE CONVERSION

Many automatic control systems are controlled by microprocessors. A microprocessor is a miniature computer, without keyboard or display devices, but with a built-in program which accepts inputs in binary digit form, performs arithmetic or logic operations on them, and outputs the result, again as a binary number.

Electrical transducers and amplifiers produce signals in the form of voltages, i.e. in analogue form (see p. 19). If these are to serve as inputs to a microprocessor, they must be converted to binary numbers by an *analogue-to-digital converter* (usually called an *A to D* or *A/D* converter).

Similarly, if a microprocessor is to control a voltage-operated or current-operated device, such as the servo valve in Fig. 9.13

(p. 185) the binary digit output must be converted into a corresponding voltage by means of a *digital-to-analogue* (*D to A* or *D/A*) *converter*.

The point in the system at which A/D or D/A conversion takes place is called the *interface*.

D/A and A/D converters are integrated circuits which require the usual 3 to 30 volt d.c. power supplies. Several types are available; one or two of them can be used for both D to A and A to D conversion, but most of them are D to A or A to D only.

Various methods have been developed for coding an analogue signal into binary digit form, but the usual method is *pulse code modulation* (*PCM* for short). This is illustrated in Fig. 3.35. The signal voltage is sampled at regular intervals, and the samples are converted into proportional binary numbers, each number corresponding to a level of voltage. In the example shown, the levels have been expressed as four-digit binary numbers, so this converter can only resolve the signal to $1/(2^4 - 1) = 1/15$ of the overall range. As can be seen, this distorts the analogue signal into a series of large steps. This distortion is called *quantization error* or *quantization noise*, since the effect is like electrical 'noise' superimposed on the original signal.

Fig. 3.35

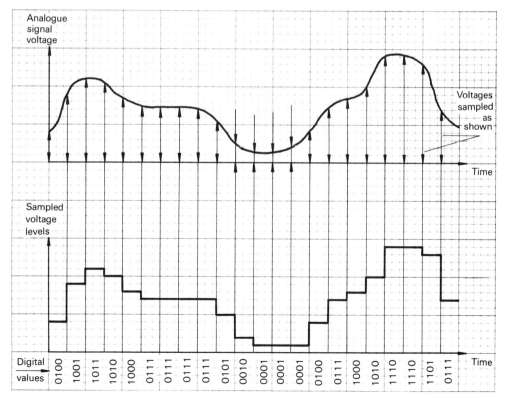

Clearly, the more binary digits (called *bits*) we can afford to use in expressing a voltage level, the more accurate our digital conversion will be — a binary number containing n bits will make each step $1/(2^n - 1)$ of the range. The practical limit at present is about 12 bits; it can be quite expensive to try to increase the resolution of the system beyond this.

Another consideration which affects the accuracy of our signal conversion is the sampling rate. It can be shown that the sampling rate must be greater than twice the maximum frequency to be reproduced in the waveform, otherwise the samples when put together will produce a different simpler waveform (called an *alias*). A further implication of this is that before the analogue signal is converted, it should be passed through a *low-pass filter* to eliminate any frequencies higher than half the sampling frequency, otherwise these will alias as lower frequencies, introducing distortion into our signal.

The stepped curve of Fig. 3.35 indicates that the sampled voltage is kept constant until the next sample is taken. If this was not done, i.e. if the voltage was allowed to vary while its A/D conversion was taking place, another error would be introduced. To keep the sample voltage constant, a sample-and-hold integrated circuit is used.

The principle is shown in Fig. 3.36. The signal enters a buffer amplifier (to keep the switching isolated from the signal source) and passes through an electronic switch to a second buffer amplifier. The switch is 'closed' for a few microseconds, to charge the capacitor C to the signal voltage, then 'opened' to hold the voltage of the sample constant. The buffer amplifiers have a voltage gain of exactly 1, so that the signal voltage is unaltered, but they have a large current gain, to maintain their output voltage constant in spite of the charging current drawn by the capacitor, and the current drawn by the A/D converter. More sophisticated circuits have means of feeding the output of the second buffer back to the capacitor, to reduce the effect of charge leakage, and hence 'voltage droop'. A/D and D/A converters have parallel outputs or inputs for the binary digits — that is, each bit in the number leaves or enters via its own particular pin.

Fig. 3.36

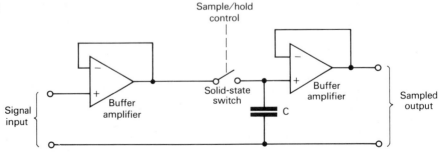

Analogue-to-digital converters are of three main types.

In the *ramp type*, a voltage increasing steadily from zero (the ramp) is continuously compared with the sample voltage. When the two become equal, a binary counter, which started counting from zero with the ramp, is stopped and its total count is output as the digital conversion. This type is the cheapest but slowest, a conversion time of about 1 ms being usual.

In the *successive approximation type*, starting at the most significant bit (MSB) and finishing at the least significant bit (LSB) — that is, the left and right hand bits of the number, respectively — each bit in turn is set to 1 or 0 as necessary, to keep the voltage represented by the digits up to then as high as possible without exceeding the sample voltage. This type is more expensive but much faster, conversion times being of the order of 20 μs.

The *flash type* of A/D converter has one comparator for each possible combination of binary digits, and it outputs the highest binary number whose voltage does not exceed the sample voltage. Since all the comparators work simultaneously, this is the fastest type, with conversion times measured in nanoseconds, but it is also the most expensive because, for example, a 6-bit flash-type converter must have $2^6 = 64$ comparators.

In **digital-to-analogue converters** each binary digit operates a solid-state switch (1 turns it on, 0 turns it off) switching currents to the summing input of an operational amplifier. Looking from the LSB to the MSB, each bit can switch on a current which is twice that of the preceding bit. The result is a current proportional to the binary number. This is converted to a proportional voltage by the amplifier and feedback resistor. Because there are no steps to go through in the conversion, D/A conversion is quite fast, the output voltage taking about 1 μs to settle.

EXERCISES ON CHAPTER 3

1 (a) State the two main types of electronic amplifier and explain the difference between them.

(b) What is an operational amplifier? Illustrate your answer with the usual circuit diagram symbol for this type of amplifier, indicating on it the power supply connections which are needed.

(c) What is meant by 'clipping' in the output of an amplifier? What is the cause of it, and how would you ensure that it does not occur?

2 (a) The compound lever mechanism of a mechanical comparator is shown diagrammatically in Fig. E3.1. Calculate the magnification of the system and the width of a scale division which would represent 1 micron (1 μm).

Fig. E3.1

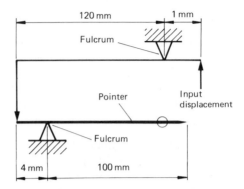

(b) Explain why the magnification obtained by a mechanical comparator does not usually exceed 5000:1, whereas magnifications of 50 000:1 can be obtained by pneumatic and electric systems.   (I.Q.A.)

3   Fig. E3.2 represents the mechanism of a mechanical comparator. Calculate the magnification of the system.

Fig. E3.2

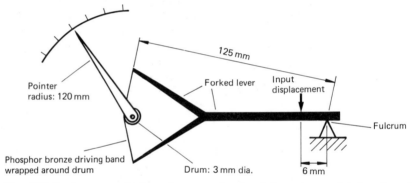

4   Fig. E3.3 shows a train of gear wheels driven by a driving band which applies a tangential pull to a drum 20 mm in diameter, carried by wheel A. The whole arrangement forms a displacement measurement system. The numbers of teeth on wheels A, B, C, D and E are 150, 75, 150, 100 and 50 respectively.

Fig. E3.3

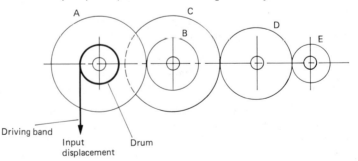

Calculate:

(a) the gain of the gear train between wheel A and wheel E;

(b) the overall gain of the system, in degrees/mm.

5 Fig. E3.4 is a diagrammatic representation of a proposed mechanism for dial test indicator. All gear teeth are module 0.3. The numbers of teeth on wheels A, C, D, E, F and G are 15, 12, 55, 12, 35 and 10 respectively.

(a) Calculate the circumferential pitch of the gear teeth (and hence the pitch of the rack teeth).

Fig. E3.4

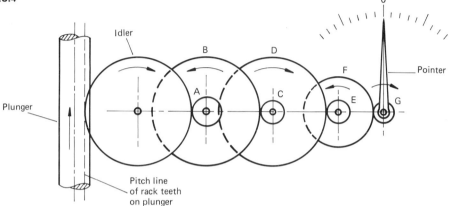

(b) It is intended that the pointer should make one complete revolution for each 0.2 mm of plunger displacement. Determine how many teeth wheel B should have, to give a plunger movement as close as possible to 0.2 mm, for one revolution of the pointer.

(c) Determine the percentage error in the instrument if wheel B has the number of teeth you have calculated in answer to part (b).

6 (a) Sketch a 'bridge' circuit.

(b) Explain why electrical comparators usually use this type of circuit.

7 (a) Make a labelled diagram of a Wheatstone bridge circuit incorporating an active strain gauge and a 'dummy' strain gauge, and show by means of a sketch how these gauges would be employed to measure the tensile force in a tie rod.

(b) In a tensile test using this arrangement, the active and dummy gauges were $200\,\Omega$ each, with gauge factor 2.15. The other two resistances in the bridge circuit were both nominally $750\,\Omega$, one of them being a variable resistor.

When the bridge was originally balanced, the resistance of the variable resistor was $748.7\,\Omega$. A tensile load was then applied to the tie rod, and the bridge re-balanced. The resistance of the variable resistor was now $751.8\,\Omega$. Calculate the strain in the tie-rod.

8 In a test to determine the modulus of elasticity of cast iron, a test specimen with a circular cross-section 20 mm in diameter was strain-gauged with a single active gauge, and loaded in tension. The

active gauge, which had a gauge factor of 2.1, was one of the four resistors of a Wheatstone bridge, all four resistances being nominally 180 Ω. The following results were obtained:

Load (kN)	0	5	10	15	20
Resistance R_B at balance (Ω)	179.84	179.88	179.92	179.96	180.02

Load (kN)	25	30	35	40	
Resistance R_B at balance (Ω)	180.06	180.10	180.16	180.20	

Determine the modulus of elasticity of the material, from a graph of the results.

9   In Fig. E3.5 R_A and R_D are strain gauges bonded to the top and bottom surfaces of a beam as shown in Fig. 3.23(b). The cross-section of the beam is a rectangle, 25 mm wide, 6 mm deep. The resistances of the gauges and fixed resistors are 120 Ω each. The gauge factor is 2.1 and the bridge energisation voltage is 15 V. The material of the beam is steel, with a modulus of elasticity of 200 GN/m².

Fig. E3.5

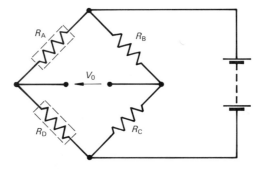

After the bridge has been balanced, the beam is loaded so that a bending moment of 40 N m is applied to the strain-gauged portion. Calculate:

(a) the stress, and

(b) the strain,

at the surfaces to which the strain gauges are bonded.

(c) the change in resistance of R_A and R_D

(d) the bridge open-circuit output voltage, V_0.

10   In a direct reading strain gauge bridge with four active arms, resistances R_A, R_B, R_C and R_D, of the basic Wheatstone bridge circuit shown in Fig. 3.18, are all active strain gauges. They are positioned and connected so that the bridge is as sensitive as

possible, i.e. so that the output voltages at P and Q change in opposite directions when the appropriate loading condition is applied to the specimen.

Fig. E3.6 shows the root of a cantilever of rectangular cross-section, with the strain gauges for three direct-reading strain gauge bridge circuits bonded to it. Each bridge circuit has four active arms. One bridge circuit is to measure shear force F_s, one is to measure direct tensile load F_t, and one is to measure bending moment M.

Fig. E3.6

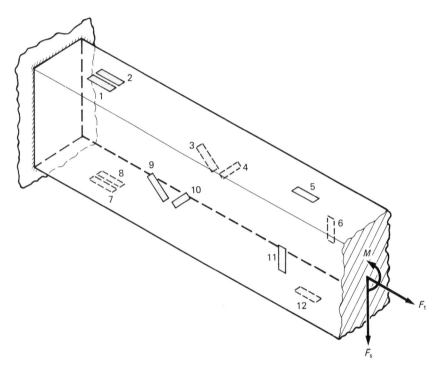

Copy the diagram below, inserting in the gap in each line an appropriate gauge number, from the gauge numbers 1 to 12 shown in Fig. E3.6.

Fig. E3.7

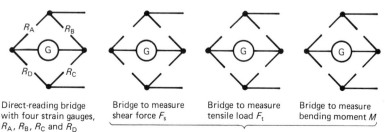

Direct-reading bridge with four strain gauges, R_A, R_B, R_C and R_D

Bridge to measure shear force F_s

Bridge to measure tensile load F_t

Bridge to measure bending moment M

Insert gauge numbers 1 to 12 from Fig. E3.6

11   (a) Draw the circuit of a resistance thermometer connected to a Wheatstone bridge using:
(i)   the three-wire system;
(ii) the four-wire system.

(b) Explain the purpose of the extra wire or wires.

12   (a) What factors govern (i) the linearity, and (ii) the resolution of a typical rotary potentiometer?

(b) A $10\,k\Omega$ wirewound potentiometer in which the wiper has $300°$ of travel, is connected to a power supply which gives constant $12\,V$ d.c. Calculate (i) the gain of the potentiometer in volts/radian, and (ii) the output voltage if a $10\,k\Omega$ load resistor is connected across the output terminals of the potentiometer when the wiper is at the mid-point of its travel. (iii) By means of a sketched graph show how the error in output voltage will vary as the wiper moves from one end of the potentiometer to the other.

13   (a) Sketch and describe the two main types of potentiometer available to a designer, distinguished from each other by the mode of movement of the slider.

(b) State three main types of material used for the resistance elements of potentiometers, and discuss the relative advantages and disadvantages of each.

14   A laboratory type of potentiometer, consisting of a resistance wire stretched along a metre rule, was used to determine the open-circuit voltage of an unknown voltage source. The voltage of the power supply to the resistance wire was first determined by connecting a standard cell of $1.0186\,V$ between the negative terminal and the slider. The balance point was shown by a galvanometer to be $25.7\,cm$ from the negative end of the wire.

The unknown voltage source was then substituted for the standard cell, and balance was found at $61.9\,cm$ from the negative end of the wire.

Calculate:

(a) the voltage of the power supply;

(b) the voltage of the unknown source.

15   A force measurement system is to consist of a coil spring, fixed at one end, and having its other end connected to the wiper of a slide-operation potentiometer (i.e. one whose track is straight, not circular). The spring has a stiffness of $50\,kN/m$, and the resistance element of the potentiometer consists of 400 turns of wire, wound uniformly on a straight insulator, forming a track $200\,mm$ long.

(a) Choose from the following the smallest change of force that can be resolved by this combination:

(i)	(ii)	(iii)	(iv)	(v)
50 N	13 N	0.025 N	25 N	100 N

(b) The resistance of the winding is 5 kΩ. If the maximum power rating of the potentiometer is 0.5 W, choose from the following, the maximum voltage which can be applied to it:

(i)	(ii)	(iii)	(iv)	(v)	(vi)
1.6 V	16 V	70 V	100 V	50 V	25 V

(c) The display device for this system is to be a voltmeter, connected between the wiper and one end of the resistance winding. What must be the minimum resistance of the voltmeter, if the error in the voltage shown, when the wiper is in mid-position, is to be less than 1% of the true voltage?

(i)	(ii)	(iii)	(iv)	(v)	(vi)
5 kΩ	25 kΩ	125 kΩ	500 kΩ	1 MΩ	5 MΩ

16 (a) What kind of transducer is a charge amplifier used with?

(b) What are the units of (i) the input to, and (ii) the output from a charge amplifier?

(c) Sketch the variation of charge amplifier output against time, labelling the graph axes appropriately, when a step input is applied to the transducer in this system.

17 (a) A piezoelectric transducer is to be used with a charge amplifier to measure a force step of about 200 N, which lasts for 2 seconds, then returns instantly to zero. If the maximum allowable error in the measurement is to be 1%, determine the minimum allowable time constant for the system.

(b) The charge amplifier has a feedback capacitor C_f in parallel with a feedback resistor R_f. The time constant of the system is given by $\tau = R_f \times C_f$, where τ, R_f and C_f are all in basic SI units. If R_f is $10^{10}\ \Omega$, determine the value of C_f to give the required time constant.

(c) The sensitivity of the piezoelectric transducer is 10 pC/N. Calculate the output voltage of the charge amplifier when the 200 N force is applied, using the value of C_f calculated in part (b).

(d) If the charge amplifier is to be followed by a simple voltage amplifier, so that the 200 N force results in a final output voltage of 5 V, what gain should the voltage amplifier have?

Chapter 4

Display Units

At the output end of any measurement system, there must be some means of displaying the measured value. The information may be displayed in digital form (that is, as an actual number), or in analogue form (that is, as a displacement proportional to the number, from which the number may be deduced). In this course, we shall study analogue display units only.

POINTER AND SCALE

A mechanically operated pointer on a scale is the simplest form of analogue display unit. The majority of instruments used for engineering measurements have this form of display — particularly the pressure gauge and the moving-coil meter, which we studied in Chapter 1.

The pointer and scale display has its disadvantages for some purposes. For instance, it leaves no record of how the reading varied with time, and its response to a change of input is fairly slow — because of its inertia, a mechanically operated pointer takes from one tenth of a second to several seconds to follow a step input to the instrument. Also it is difficult to compare values on two or more instruments which have rapidly changing inputs. If any of these disadvantages is serious, we must use one of the display devices discussed later in this chapter. But for most measurement systems a pointer and scale display is perfectly adequate.

Apart from poor illumination, the most likely source of error, when reading from a pointer and scale display, is *parallax*. This is an error which occurs when the line of sight is not normal to (i.e. not perpendicular to) the dial face, so that the reader views the pointer against a point on the scale to one side of its true position. To reduce this error, the clearance between the scale and the pointer should be as small as possible. To eliminate it, some scales have an arc of mirror set into the scale, just inside the boundary line of the scale. For accurate reading, the reader moves his head so that the pointer and its reflection in the mirror coincide — he can then be sure that his line of sight is normal to the scale.

OPTICAL POINTERS

To reduce the inertia of the mechanical pointer, an 'optical pointer' is used where high-speed response with minimum overshoot are required. The principle is shown in Fig. 4.1. The rotating part of the measuring instrument (for example, the coil of a moving coil meter) carries a very small mirror. The 'pointer' is a ray of light projected on to the mirror from a fixed lamp, and reflected back on to a graduated scale. The lamp and the scale are at different levels, above and below the mirror, so that the reflected ray is not interrupted by the lamp.

Fig. 4.1 Principle of the optical pointer

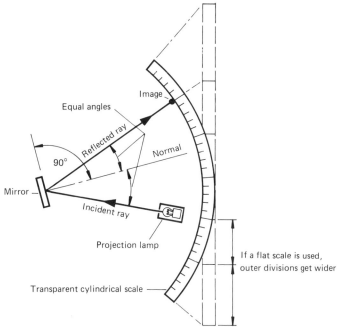

Apart from the smaller moment of inertia of a tiny mirror compared with a long pointer, the optical pointer has two other advantages:

(a) The scale can be positioned at any distance from the mirror: the greater the distance, the more sensitive the instrument, with no increase in moment of inertia.

(b) The optical pointer rotates through twice the angle that the mirror rotates through. This is because the reflected ray and the ray from the lamp are always at equal angles to the normal. (The normal is an imaginary line perpendicular to the mirror at the point of reflection.) Thus if the mirror (and hence the normal) rotates through an angle of $\alpha°$, the reflected ray rotates through $2\alpha°$. So an optical pointer instrument has twice the sensitivity of a mechanical pointer of the same length.

What are the disadvantages? Well it is possible to make an instrument too sensitive, so that the image judders violently at the slightest disturbance. So we must be careful not to overdo this sensitivity thing. Also the optical system is more expensive and more complicated than a simple pointer. And it is more easily disturbed, so, while it is good for comparative measurements, it is more likely to be subject to zero error than a simple pointer. A further disadvantage is that the reflected image is usually more difficult to see than a mechanical pointer. However, in some applications an optical pointer can be extremely useful, as we shall see when we study the ultraviolet recorder.

THE ULTRAVIOLET RECORDER

The ultraviolet recorder (usually referred to by its initials as the *UV recorder*), is the most sophisticated recording instrument which has so far been devised. In the present state of the art, an ultraviolet recorder can record, in the form of amplitude/time graphs, up to 50 independent channels of data, on a roll of paper 305 mm wide, unwinding at controlled speeds of up to 4 m/s. And its sensitivity can remain practically constant over the frequency range from 0 to 12 kHz. Higher frequencies can be recorded on a magnetic tape recorder, but they cannot be displayed by it. Higher frequencies can be displayed on an oscilloscope but they cannot be permanently recorded by it.

How does this marvellous instrument work? The essential transducer is our old friend, the moving-coil meter from Chapter 1, but here it is refined into a galvanometer as slim as a pencil, the moving coil suspended between bobbins and pulled taut by a suspension wire above and below, which also acts as the electrical connection to the coil. And this suspension wire carries a tiny mirror instead of a pointer, to reflect a ray of ultraviolet light which does the recording. Fig. 4.2 shows a typical galvanometer for a UV recorder, and Fig. 4.3 illustrates its construction.

The ultraviolet light is emitted by a mercury-vapour lamp, and directed by an optical system on to the mirrors of all the galvanometers in the recorder. Each of these reflects its own ray of ultraviolet light on to a mirror-and-cylindrical-lens strip, positioned across the full width of the paper. This focuses the rays from the galvanometers into spots of ultraviolet light on the recording paper. Fig. 4.4 shows the main features of a UV recorder.

The recording paper carries an emulsion sensitive to ultraviolet light, and as it unrolls, the spots focused on to it from the galvanometer mirrors draw 'graphs' of current or voltage (measured across the paper) against time (measured along the paper). These 'graphs' (called *traces*) are not immediately visible, but appear after a few

Fig. 4.2 Typical galvanometer. *Source:* Southern Measuring Instruments Ltd, Sandown, I.O.W.

Fig. 4.3 Diagram of the construction of a typical galvanometer. *Source:* Southern Measuring Instruments Ltd, Sandown, I.O.W.

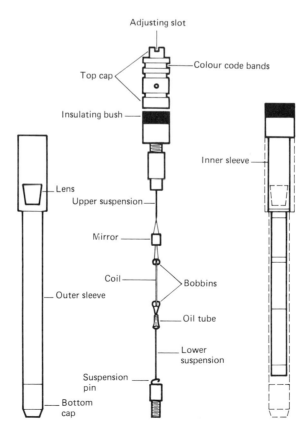

Adjusting slot

Colour code bands

Top cap

Insulating bush

Lens

Upper suspension

Mirror

Coil

Bobbins

Oil tube

Lower suspension

Suspension pin

Outer sleeve

Bottom cap

Inner sleeve

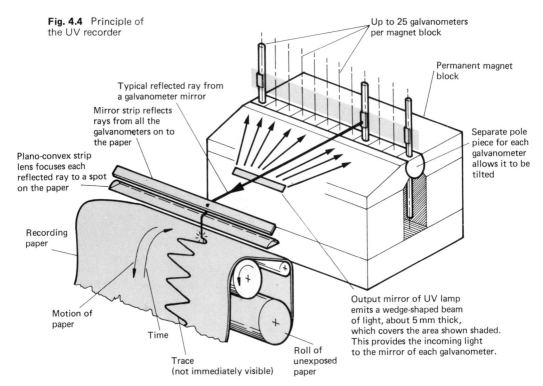

Fig. 4.4 Principle of the UV recorder

Up to 25 galvanometers per magnet block

Permanent magnet block

Typical reflected ray from a galvanometer mirror

Mirror strip reflects rays from all the galvanometers on to the paper

Separate pole piece for each galvanometer allows it to be tilted

Plano-convex strip lens focuses each reflected ray to a spot on the paper

Recording paper

Motion of paper

Time

Trace (not immediately visible)

Roll of unexposed paper

Output mirror of UV lamp emits a wedge-shaped beam of light, about 5 mm thick, which covers the area shown shaded. This provides the incoming light to the mirror of each galvanometer.

seconds, developed automatically by the small proportion of ultraviolet light present in normal room lighting. This same ultraviolet light will darken the rest of the paper in a matter of weeks, so that the traces gradually disappear into the background, but if a permanent record of the traces is required, the paper can be chemically treated to stabilise it, before this happens.

Further systems of lenses and mirrors focus light from the mercury vapour lamp to draw grid lines along the paper, and timing lines across the paper, to make it easier to measure current (or voltage) and time from the trace, and each trace in turn is interrupted automatically, so that the traces can be identified one from another.

Because the 'pointer' of each galvanometer is a ray of light, which has no mass, and therefore no moment of inertia, very high frequencies can be recorded by means of a UV recorder. Also, since the rays of light can pass through each other unhindered, each trace can traverse the full width of the paper without interfering with any other trace — which is why trace identification is such a necessary feature of a UV recorder.

THE XY PLOTTER

The XY plotter, a typical example of which is shown in Fig. 4.5 is an analogue display device which draws a graph to show the relationship between two signal voltages. One voltage, applied to the X-input terminals, displaces the pen a proportional distance from left to right across the paper; the other voltage applied to the Y-input terminals, displaces the pen a proportional distance from 'bottom' to 'top' of the paper.

Fig. 4.5 XY plotter.
Source: Gould Advance
Ltd, Hainault, Essex

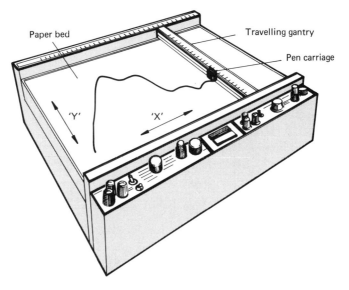

The XY plotter can be set up to plot its output curve on a graph paper sheet: by adjusting the zeroing control for each signal channel, the pen can be positioned so that it starts from the origin (the intersection of the axes) wherever it may be on the paper, and if the gain of each channel is set to one of the standard sensitivity values (V/mm) available on the instrument, the axes can be graduated in values of input signal voltage.

A typical application of an XY plotter would be the plotting of a stress–strain curve from a tensile test.

Because the pen has to move as fast as possible if it is to follow rapidly changing signal voltages, it has to be driven by two electric motors — one moving the pen along the X-axis, the other moving it along the Y-axis. Each motor is driven by its own electrical servo, the input to which is the signal voltage input to that channel. Fig. 4.6 is a block diagram of the servo, electric motor and feedback potentiometer for one of the two channels; the diagram for the other channel would be identical. The symbol  represents an arrangement of resistances which adds together the input signal voltage and the voltage from the zeroing potentiometer and subtracts from the sum a voltage proportional to the displacement of the pen, obtained from the feedback potentiometer. The resultant of these three voltages goes to one input of an operational amplifier (the pre-amplifier). The output of the pre-amplifier goes as input to the power amplifier, and the output of the power amplifier forms the power supply to the electric motor. The electric motor rotates, through a reduction gear train, the drive shaft carrying both the cable drive pulley and the wiper of the feedback potentiometer.

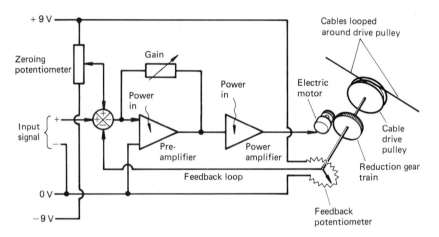

Fig. 4.6   Block diagram of the X- or Y-channel, showing how the input signal is converted into a proportionate displacement of the cable driving the pen

The effect of the feedback loop is that the electric motor rotates the drive shaft to a position where the voltage fed back from the wiper of the feedback potentiometer exactly cancels out the sum of the zeroing and signal voltages. This is an example of the automatic control circuits we shall meet in Chapter 8.

Fig. 4.7 shows one way of converting the rotation of the X- and Y-axis servo motors into displacement of the pen.

The arrows show the direction of movement of the cables for a clockwise rotation of the X- or Y-axis drive shaft. The two movements are entirely independent of each other, the Y-cables just idling around the pen carriage pulleys when the gantry is in motion.

An alternative design has the servo motors mounted on the gantry, and connected to the amplifiers by flexible wiring. This results in a much simpler system of interconnecting cables and pulleys, and hence less friction, but increases the inertia of the gantry.

Inertia and friction are the factors which limit the performance of an XY plotter. The performance of an XY plotter is mainly specified in terms of writing speed or slewing speed, and maximum acceleration of the pen. *Writing speed* is the maximum pen speed

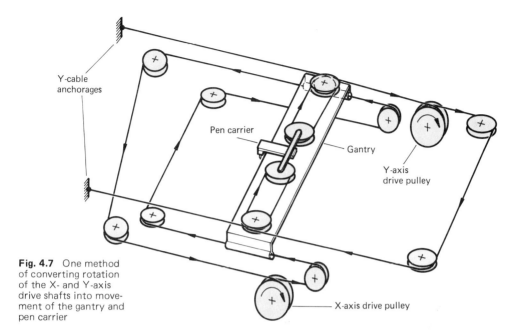

Y-cable anchorages

Pen carrier

Gantry

Y-axis drive pulley

Fig. 4.7 One method of converting rotation of the X- and Y-axis drive shafts into movement of the gantry and pen carrier

X-axis drive pulley

attainable along either axis. In American specifications this is called *slewing speed*; in British specifications, however, slewing speed is the vector sum of the writing speeds along each axis, i.e.

Slewing speed

$$= \sqrt{(\text{max. speed along X-axis})^2 + (\text{max. speed along Y-axis})^2}$$

A good XY plotter has a writing speed better than 0.5 m/s and can accelerate the pen at more than 20 m/s².

These figures limit the frequency response of the XY plotter. At an amplitude equal to half the width of the paper, the plotter can accurately follow sinusoidal a.c. frequencies up to only about 0.5 Hz. At frequencies above this, the amplitude of the displayed sine wave must be reduced, to enable the pen to keep up with the input signal. The upper limit is about 10 Hz, at which frequency the amplitude of the displayed sine wave has to be very small indeed. So the XY plotter is strictly for very low frequencies.

The XY/t plotter. Most XY plotters have an optional time base which can be switched in, or plugged in, which applies a steadily increasing voltage to the X-servo, driving the pen at a constant speed from left to right. These instruments are called XY/t plotters. Using the time base converts the XY plotter into a Y–t plotter, which draws a graph of voltage on the Y-input against time. This makes its display similar to that of a cathode-ray oscilloscope, the difference being that the plotter's display is permanent but limited to very low frequencies.

THE CATHODE-RAY OSCILLOSCOPE

The cathode-ray oscilloscope is a display device which is probably well known already to the reader. It consists of a cathode ray tube together with various electronic circuits which may be represented by the blocks in Fig. 4.8.

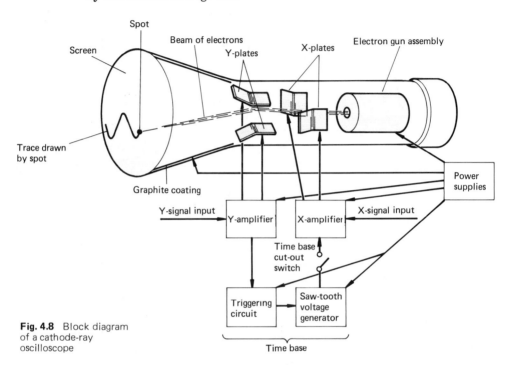

Fig. 4.8 Block diagram of a cathode-ray oscilloscope

The cathode ray tube is a glass bulb with an enlarged end which is flattened to form a screen. Inside the tube is a complete vacuum. The screen is coated on the inside with a phosphorus-based material (the *phosphor*) which glows wherever a beam of electrons strikes it. The electrons are emitted by a heated cathode, and are accelerated to a very high speed (some millions of metres per second) by the attraction of the various anodes through which they pass. The final anode consists of the screen itself, which is lightly aluminised to make it conductive, and connected to a graphite coating inside the conical end of the tube. This is at about + 2000 V relative to the cathode.

The heater, cathode, control grid and three anodes which accelerate and focus the beam of electrons are constructed as an assembly called the *electron gun*. This assembly also carries the two pairs of beam deflector plates: the X-plates and the Y-plates.

The deflector plates are arranged in pairs so that the beam of electrons passes between each pair. The X-plates are vertical and deflect the beam horizontally; the Y-plates are horizontal and deflect the beam vertically. This deflection is produced by equal and opposite voltages applied to the X-plates by the X-amplifier and to the Y-plates by the Y-amplifier. The electron beam, being a stream of negative charges, bends closer to the plate with the positive voltage, and away from the plate with the negative voltage.

The oscilloscope is normally used to display a 'graph' of voltage against time, as shown in Fig. 4.9. The voltage is the signal input to the Y-amplifier. This is amplified to produce equal and opposite voltages on the Y-plates. It also 'fires' the trigger circuit, causing the saw-tooth voltage generator to produce a voltage which steadily increases to a maximum, then quickly returns to zero. This voltage, applied to the X-amplifier (the time base cut-out must be closed), produces equal and opposite voltages on the X-plates, which deflect the electron beam steadily from left to right across the screen, then return it very quickly (*fly-back*) to its starting point.

Fig. 4.9 Normal oscilloscope display

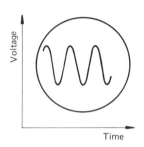

It should be noted that although in principle either pair of deflector plates could act as Y-plates (provided that the tube was rotated to bring them horizontal), and Fig. 4.8 shows them as the pair nearer the screen to simplify the diagram, in fact they are usually the pair further from the screen, as this gives greater Y-sensitivity.

SELF-TEST QUESTION 8 (Solution on p. 236)

Why is greater sensitivity of the Y-channel obtained by using the deflector plates further from the screen as the Y-plates?

USING THE OSCILLOSCOPE AS AN XY DISPLAY DEVICE

Just as an XY plotter can produce a voltage/time graph if the X-servo is operated by a time base circuit, so an oscilloscope can produce an XY display if the time base is cut out and a second signal is input to the X-amplifier as shown in Fig. 4.8.

A typical application occurs where an oscilloscope is used as the display device in an engine indicator system. An engine indicator gives a pressure/volume graph showing the pressure variation in an engine cylinder during one complete engine cycle. Originally engine indicators were all-mechanical devices, but such instruments cannot be used on high-speed engines as the inertia of their moving parts is too great. A cathode-ray engine indicator, however, has no speed limitation at all.

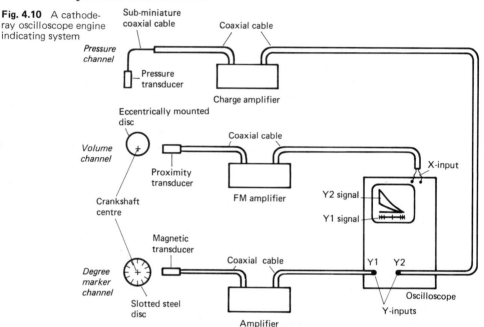

Fig. 4.10 A cathode-ray oscilloscope engine indicating system

Fig. 4.10 shows the equipment needed for a complete cathode-ray oscilloscope engine indicating system. The oscilloscope time base is switched off and an inductive transducer, operated by an eccentric circular disc on the engine crankshaft, produces an X-input signal proportional to the displacement of the piston. The pressure in the engine cylinder is sensed by a piezoelectric pressure transducer connected to one of the Y-inputs of the oscilloscope. This, in conjunction with the X-input, gives the engine indicator diagram shown in Fig. 4.11.

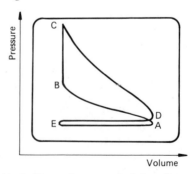

Fig. 4.11 Cathode-ray oscilloscope engine indicator display: AB, compression stroke; BC, firing; CD, power stroke; DE, exhaust stroke; EA, induction stroke

The slots of a slotted disc, also carried by the crankshaft, induce pulses in a magnetic transducer, and these are fed to the other Y-input of the oscilloscope, where, in conjunction with the X-input, they produce a horizontal trace marked with 5°, 10°, 20°, etc. of crankshaft rotation, for projection on to the indicator diagram.

COMPARISON OF DISPLAY DEVICES

The following table summarises the main characteristics of the various display devices we have considered up to now.

Display device	Typical response time to a step input	Typical bandwidth (frequency range of a.c. which can be accurately displayed)	Mains power required?	Remarks
Moving-coil meter	About 0.5 seconds (depends on the damping of the meter movement)	—	No	Pointer and scale display
UV recorder	About 100 μs (100×10^{-6} seconds)	0–12 kHz (depends on the damping of the galvanometers)	Yes	Semi-permanent display on continuous strip of sensitised paper.
XY plotter	About 0.3 seconds (depends on writing speed)	0–0.5 Hz at 190 mm amplitude; 0–5 Hz at 12 mm amplitude	Yes	Permanent display of two signal voltages on X and Y axes on standard size paper such as A3. The instrument is usually convertible to a Y–t plotter by using an internal time base.
Cathode-ray oscilloscope	10 ns (10×10^{-9} seconds)	0–30 MHz	Yes	Transient display of voltage/time on screen of cathode-ray tube, though special purpose oscilloscopes can give long-persistence display. Convertible to XY display if time base switched off. Double-beam oscilloscope gives simultaneous plot of two signals against time.
Mechanically operated pointers	Depends on length of pointer, mechanical strength of pivots, power available to operate pointer, inertia of pointer and mechanism, etc. A response time of 10 ms should be obtainable with careful design.	—	—	—

EXERCISES ON CHAPTER 4

1 (a) By means of a simple labelled diagram, show how a ray of light may be used in place of a mechanical pointer on an instrument such as a moving coil galvanometer.

(b) Explain the advantages and disadvantages of such an arrangement.

2 What is the effect known as parallax, and what design features are usually employed to avoid it?

3 (a) Explain how a UV recorder produces a record of the variation with time of an electrical signal.

(b) Does it produce a permanent record? If not, what must be done to make it permanent?

(c) How is time measured on the record, and how is a datum obtained, from which to measure the displacement of the trace?

(d) How are the traces of the different signals which may appear on the same record identified, one from another?

(e) What limits the displacement of any trace from its datum?

4 (a) Make a sketch of an XY plotter, and indicate on it the gantry and the pen carriage. Show, also, the respective directions in which the X and Y signals are measured.

(b) Explain how the incoming X signal or Y signal is converted into a proportionate displacement of the recording pen. Illustrate your answer with a simple block diagram.

(c) What are the two alternative ways of applying displacements to the pen carriage and the gantry? Explain the advantage and disadvantage of each method.

(d) Explain what is meant by the terms (i) writing speed, and (ii) slewing speed, in the specification of a British XY plotter.

(e) Suppose that you have to record, as accurately as possible, on an XY plotter, a waveform with a frequency of about 4 Hz. How would you ensure that the writing speed was not exceeded?

(f) What additional device is needed to convert an XY plotter into an XY/t plotter, and what does it do?

5 (a) Make a diagram of a cathode-ray tube, indicating on it the electron gun, the X and Y plates, the beam of electrons, the screen, with a partly completed trace, and the graphite coating.

(b) Turn your diagram for part (a) into a block diagram for a complete cathode-ray oscilloscope. There should be a block each for (i) power supplies, (ii) X amplifier, (iii) Y amplifier, (iv) triggering circuit and (v) saw-tooth voltage generator. Indicate, by means of arrows, the relationships between the various blocks, the components of the cathode-ray tube, the Y input signal and a possible X input signal.

(c) When a trace is displayed on a cathode-ray oscilloscope screen in normal use, what do (i) vertical, and (ii) horizontal measurements correspond to?

(d) What action is necessary to convert a normal cathode-ray oscilloscope into an XY display device?

(e) State a typical measurement application for which a cathode-ray oscilloscope would be used as an XY display device.

6   Choose, from the following, the value nearest to the response time to a step input which you would expect from (a) a moving-coil meter, (b) a UV recorder, (c) an XY plotter and (d) a cathode-ray oscilloscope:

(i)	(ii)	(iii)	(iv)	(v)	(vi)	(vii)
100 s	1 s	10 ms	100 μs	1 μs	10 ns	100 ps

7   Choose, from the following, the value nearest to the frequency limit for the accurate recording or displaying of a sinewave on:

(a) a UV recorder,   (b) an XY plotter and   (c) a cathode-ray oscilloscope:

(i)	(ii)	(iii)	(iv)	(v)	(vi)	(vii)
0.01 Hz	1 Hz	100 Hz	10 kHz	1 MHz	30 MHz	1 GHz

Chapter 5

Measurements of Frequency, Displacement, Force and Pressure

FREQUENCY MEASUREMENT

The most general definition of *frequency* is *rate of repetition of a periodic disturbance measured in cycles/second or hertz*. In this section we shall have to limit our consideration of frequency measurement to the measurement of rotational frequency (i.e. speed of rotation in revolutions per minute), though where the device we are considering could also be used to measure the frequency of mechanical vibrations, we shall take note of that fact.

In spite of the SI system, which decreed that the basic unit of time shall be the second, engineers still measure rotational speeds in revolutions per minute. (The approved abbreviation is *rev/min*, not r.p.m.) Vibrations, however, are measured in hertz (cycles per second). This may seem illogical, but at least it tells us whether people are talking about a speed or a vibration.

THE EDDY-CURRENT TACHOMETER

A tachometer is used to measure rotational speed. The eddy-current tachometer is shown in Fig. 5.1. The shaft whose speed is to be measured is made to drive a U-shaped magnet which rotates inside a soft iron casing. An aluminium cup is suspended between the magnet and the iron casing so that the field of the magnet passes through the aluminium. As the magnet rotates relative to the cup, its magnetic field generates a current (the *eddy current*) which circulates through the material of the cup. This current sets up its own magnetic field which is attracted to that of the rotating magnet, so that the cup tends to be carried round by a magnetic torque which is proportional to the eddy current, which, in turn, is proportional to the speed of the magnet.

The cup is carried by a spindle which is free to rotate through part of a revolution. It rotates to a position in which the magnetic torque on the cup is exactly balanced by the torque from a torsion spring opposing the rotation. At its end, the spindle carries a pointer which indicates the speed on a dial.

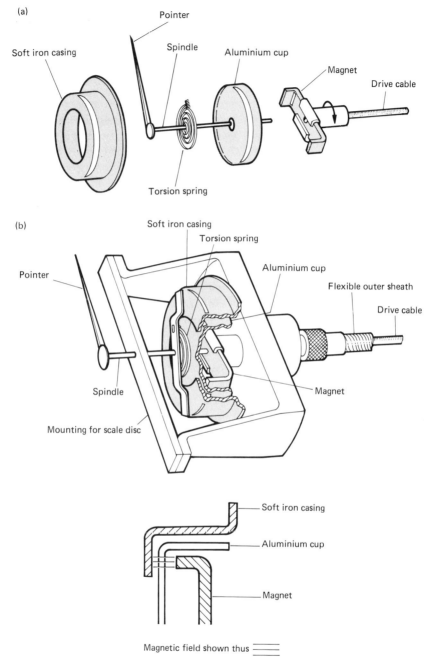

Fig. 5.1 (a) Essential components of an eddy-current tachometer. (b) Cutaway view of a typical eddy-current tachometer assembly. (c) Magnetic field which generates the eddy current

This device forms the basis of a hand-held tachometer. The drive to the magnet is by friction, through a rubber cone which is pressed into a conical centre-hole in the shaft whose speed is to be measured. The drive is transmitted through a miniature gearbox so that the tachometer can be used over several speed ranges. The gearbox also incorporates an automatic reversing gear, in the form of an idler pinion which is brought into engagement by friction when the rubber cone is rotated in the 'wrong' direction. This ensures that the magnet always rotates in the forward direction.

The eddy-current tachometer is also used as a speedometer in most motor cars. In this application, the magnet is driven by a steel cable rotated by a gear driven off the final drive shaft of the car's gearbox. Thus it measures road speed by measuring the rotational speed of the road wheels, but the dial is graduated in miles or kilometres per hour, instead of rev/min.

THE TACHOGENERATOR

Mechanical tachometers such as the eddy-current tachometer cannot be used as remote-reading instruments, as they do not produce an electrical signal which could be transmitted over long distances, and if a long flexible driving cable is used, it wastes power in friction and gives rise to stability and vibration problems. So for the remote indication of rotational speed, a tachogenerator is usually used. This is a very simple system — the transducer is a little d.c. generator (a dynamo) with its armature rotating in a permanent-magnet field. Its electrical output is connected to a voltmeter graduated in rev/min. It works because the output voltage of such a generator is directly proportional to its rotational speed, provided that very little current is taken from it. Fig. 5.2 illustrates the system.

The tachogenerator is also a very convenient transducer to use in automatic speed control systems: the generator output voltage can be fed back and compared with a set voltage, the difference (the error signal) causing the speed to be restored to the set value.

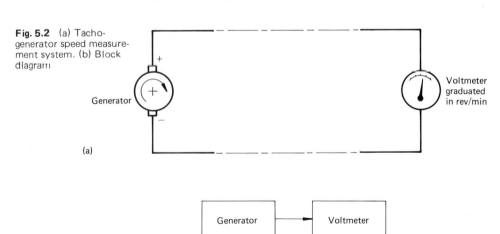

Fig. 5.2 (a) Tacho-generator speed measure-ment system. (b) Block diagram

THE STROBOSCOPE

If a rotating shaft or wheel carries a radial mark as shown in Fig. 5.3, and is illuminated by a flash or pulse of light each time the mark arrives at the same position, the mark will appear stationary. If we then know the rate of flashing (i.e. the number of light pulses per minute) we know the speed of the shaft or wheel in rev/min, as they are numerically equal.

Fig. 5.3 Stroboscope

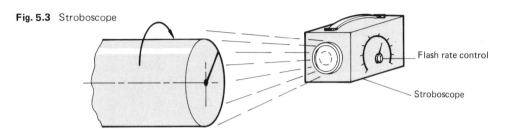

Flash rate control

Stroboscope

The piece of equipment which produces pulses of light at a controllable frequency is called a *stroboscope*, or *stroboflash*. It is illustrated in Fig. 5.3. The pulses of light are produced by a neon or xenon lamp (a filament lamp cannot be switched on and off quickly enough). The frequency of flashing is set by a pointer-knob which also indicates the frequency on a circular scale.

Thus, to measure a rotational speed, we shine a stroboscope on to a marked wheel or shaft, adjust the flash speed control until we get a stationary image, and read the speed from the control knob.

There is a snag. We shall also get a stationary image if the mark is illuminated once every two revolutions, or every three revolutions, or every four revolutions, and so on. So we are liable to get a reading of a half, or a third, or a quarter, etc., of the speed and think we have the true speed. We must always be on guard against this when using a stroboscope.

One way to avoid this error is to have some idea of the true speed already, by using some other speed measuring device as a check. But when we have to rely on the stroboscope alone for speed measurement, the following procedure should always be used. It also enables the stroboscope to be used to measure rotational speeds greater than the highest speed of flashing obtainable from it.

Absolute Speed Measurement Using a Stroboscope

With the stroboscope aimed at the marked end of the shaft, its flash rate is slowly and steadily reduced, starting with the highest rate and working steadily through the speed range. Each time a stationary radial line appears, the corresponding flash speed is noted. Various other stationary patterns of double, triple and multiple lines will appear at other speeds; these should be ignored — it is only the single image we are interested in.

At the end of this procedure we shall have a list of speeds. The true speed of the shaft can be determined by comparing successive speed ratios with the list of the theoretical speed ratios. The theoretical ratios are:

$$\frac{\text{Full speed}}{\frac{1}{2}\text{ speed}}, \quad \frac{\frac{1}{2}\text{ speed}}{\frac{1}{3}\text{ speed}}, \quad \frac{\frac{1}{3}\text{ speed}}{\frac{1}{4}\text{ speed}}, \quad \frac{\frac{1}{4}\text{ speed}}{\frac{1}{5}\text{ speed}}, \quad \text{etc.}$$

i.e. the theoretical ratio values are 2, 1.5, 1.33, 1.25, etc. The first five speed values are usually all that is needed to establish the value of the true speed beyond any doubt.

EXAMPLE   By means of a stroboscope, a stationary single image was obtained at flash speeds of 5900, 3050, 1995, 1505 and 1203 flashes per minute. Determine the true speed of the shaft.

SOLUTION   We can draw up the following table:

Ratio of successive measured speeds	*Ratio of successive theoretical speeds*
$\dfrac{5900}{3050} = 1.93$	$1 \div \tfrac{1}{2} = 2$
$\dfrac{3050}{1995} = 1.53$	$\tfrac{1}{2} \div \tfrac{1}{3} = 1.5$
$\dfrac{1995}{1505} = 1.33$	$\tfrac{1}{3} \div \tfrac{1}{4} = 1.33$
$\dfrac{1505}{1203} = 1.25$	$\tfrac{1}{4} \div \tfrac{1}{5} = 1.25$

Comparing the values of the ratios in the two columns we can see that they agree approximately, allowing for slight errors of reading and instrument calibration. We can therefore be sure that the true speed is approximately 5900 rev/min.

A more accurate value can be obtained by averaging the appropriate multiples of all the readings obtained, as below:

$$\text{True speed} = \frac{5900 + 2 \times 3050 + 3 \times 1995 + 4 \times 1505 + 5 \times 1203}{\text{Number of readings (5)}}$$

$$= \frac{5900 + 6100 + 5985 + 6020 + 6015}{5}$$

$$= \frac{30\,020}{5}$$

$$= 6004 \text{ rev/min}$$

SELF-TEST QUESTION 9 (Solution on p. 236)

To determine the speed of a compressed-air driven grinder, a radial line was painted on the grinding wheel and it was illuminated by a stroboscope while it was running at steady speed. Working down from the maximum flash frequency, a stationary image was obtained at frequencies of 5050, 3850, 3000, 2525 and 2200 flashes per minute. Estimate the speed of the grinder.

Other Uses of the Stroboscope

The stroboscope measures rotational speed by 'freezing' the movement so that the rotating body appears stationary. It can 'freeze' a vibration in the same way, so that, provided the amplitude of the vibration is large enough to allow it to be seen clearly, we can use a stroboscope to measure its frequency. Again we shall have to be careful to get the true frequency, and not some fraction of it, so we shall have to use the same procedure as for rotational speed measurement with the stroboscope.

This property of apparently stopping movement makes the stroboscope a useful instrument for examining the behaviour of a mechanical process while it is moving too fast to be seen clearly in the ordinary way. Such effects as valve bounce, the 'whirling' of shafts, or the turbulence of the air flow over fan blades (shown by tufts of wool stuck to the surface) can be studied in detail while the parts are moving at high speed, and cyclic variations can be observed in slow motion by setting the stroboscope flash speed to a slightly different frequency from the frequency of the cycle.

There is a low-speed limit at which the eye ceases to be deceived by the stroboscopic effect. This limit varies for different people; it also depends on the relative increase of illumination caused by the flash, but it is usually in the range 200–300 flashes per minute. It can therefore be taken that the stroboscope is unusable at frequencies below 200 flashes per minute.

THE PULSE COUNTER

Pulse counting is usually a function of an electronic instrument called a *timer-counter*. Fig. 5.4 is a typical block diagram of such an instrument. The essential feature is the 'clock'. This is a crystal-controlled oscillator which gives an a.c. output of square waveform at constant frequency. The crystal which controls the oscillator is a piezoelectric crystal, in the form of a slice of quartz machined to such a thickness that its natural frequency of mechanical vibration is approximately the required clock frequency. When incorporated into an electronic oscillator circuit as an inductive component, it resonates electrically at approximately the same frequency as its natural frequency of mechanical vibration. The actual frequency is adjusted by a trimming capacitor.

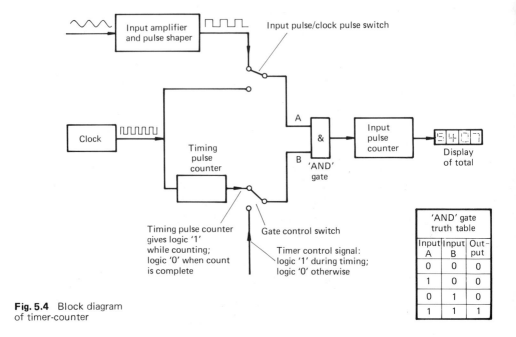

Fig. 5.4 Block diagram of timer-counter

Since the natural frequency of the crystal depends only on its dimensions and physical properties, a crystal-controlled oscillator keeps its frequency constant to within a few millionths of the required value. Hence its use as the 'clock' in the timer counter.

Fig. 5.4 shows the timer-counter system switched to the measurement of frequency. In this mode, the clock is used to time one second with great accuracy; this is done by counting its output pulses in the timing pulse counter. If the clock frequency is 100 kHz (a typical frequency for such an instrument), the timing pulse counter will give a logic signal of '1' while it is counting from 0 to 100 000 and a logic signal of '0' afterwards. The logic signal of '1' while applied to input B of the AND gate, 'holds the gate open', allowing pulses to pass through input A and be counted in the input pulse counter. These other pulses are cycles of the unknown frequency being measured. They have been shaped into squarewave pulses by the input amplifier and pulse shaper.

When the output of the timing pulse counter changes from logic '1' to logic '0' at input B of the AND gate, no more pulses can get through input A, and the display shows the total number of pulses of the unknown frequency counted by the input pulse counter during one second — that is, the frequency in hertz.

It should be noted that although the second may have been timed to an accuracy of one part in a hundred thousand, the input pulse counter can only count complete cycles of the unknown frequency — it cannot count the odd part of a cycle. This may give rise to a considerable percentage error when measuring low frequencies. More sophisticated instruments can overcome this by measuring the *period* of an unknown frequency, by timing a number of cycles, and then obtaining the frequency as the reciprocal of the period.

To use the instrument shown in Fig. 5.4 as a timer, the two switches must be switched over to their alternative positions. This admits clock pulses to the input pulse counter through input A of the AND gate, as long as there is a logic signal of '1' on input B. This logic signal of '1' is generated by an electrical pulse applied to the timing control terminals of the instrument to start it timing, and continues until a further electrical pulse generates logic '0' to stop the timer. The display then shows the number of clock pulses which have been counted between the two timing control pulses — i.e. the time interval in units of 10 microseconds, if the clock frequency is 100 kHz.

Rotational Speed Measurement by Pulse Counter

Fig. 5.5 shows a simple circuit which uses the pulse counter described above to give a digital read-out of rotational speed in revolutions per minute. This speed measurement can be repeated every two or three seconds and the value displayed is accurate to within 1 rev/min.

The three components of this system are a 60-toothed mild steel wheel, a magnetic transducer, and the pulse counter.

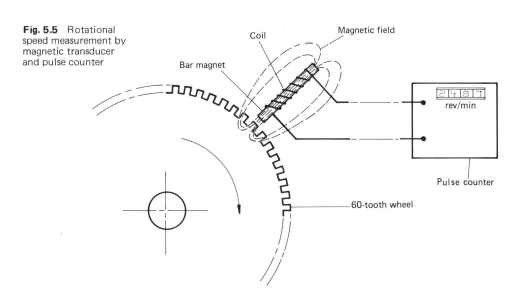

Fig. 5.5 Rotational speed measurement by magnetic transducer and pulse counter

The 60-toothed wheel could be a gear wheel, or just a mild steel disc milled with 60 slots equally spaced around its circumference. It is carried by the shaft whose speed is to be measured (so this system is really only suitable for permanent installation).

The magnetic transducer is a bar magnet with a coil of wire wound on to it. The ends of the coil are connected to the signal input of the pulse counter. As a tooth on the wheel approaches the end of the bar magnet, the steel of the tooth provides an easier path for the magnetic field than the gap between teeth did, and this increases the flux density of the magnetic field (i.e. increases the number of complete loops of magnetic field like those shown in Fig. 5.5). This increase of magnetic flux density along the axis of the coil induces a voltage in the coil proportional to the rate of increase. The same effect occurs when the tooth has passed the centre of the bar magnet and is going away from it, only in this case the voltage is induced in the opposite direction. The total effect is that the magnetic transducer generates an a.c. voltage with the same number of cycles per second as the number of teeth passing the magnet per second.

The pulse counter converts the a.c. cycles into square wave cycles, counts them for exactly one second, and displays the number counted. Thus, with a 60-tooth wheel the pulse counter displays:

Cycles per second

> = Number of teeth passing the magnet per second
>
> = 60 × Revolutions per second
>
> = Revolutions per minute

We seem to have here a perfect way of measuring rotational speed. Are there any snags? Well, a simple pulse counter cannot count part of a pulse, so it will chop off any decimal fraction of a rev/min. This might cause an appreciable percentage error at very low speeds.

The system could also *over*-estimate the speed if the connecting wires, from transducer to counter, pick up any additional pulses — perhaps from electrical contacts which are 'sparking' close by. To make sure this does not happen we should use coaxial cable, which is self-screening, for the connecting leads.

As long as we bear in mind the above points, the system of toothed wheel, magnetic transducer and counter provides a very reliable and accurate method of measuring rotational speed.

DISPLACEMENT-MEASURING DEVICES

We have already considered one displacement-measuring device, the linear variable differential transformer. In this section we shall consider one more: *the float.*

THE FLOAT

The float is a simple method of measuring the displacement of the surface of a liquid — in other words it is a 'water-level'-indicating device. We can see an example in a toilet cistern: the ball of the ball-cock is a float which operates a lever to close off the inflow of water from the water-main as the water-level approaches the required height. This, however, is more an automatic control system than a measurement system. For an example of a float in a measurement system we can consider the fuel-level unit, in the fuel tank of a motor car. Fig. 5.6 shows the principle on which it works.

The float is on the end of an arm, pivoted near the top of the tank. As the level in the tank changes, the float rotates the arm through a small angle, and this rotates the 'wiper' — a springy copper strip which slides over a winding of resistance wire, earthing the point

Fig. 5.6 (a) Fuel level, transmitter unit. (b) Circuit diagram of system

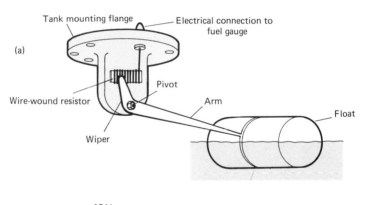

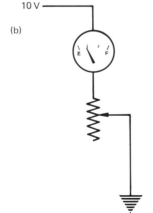

at which it makes contact. Thus the tank unit of the fuel gauge system provides a variable-resistance path to earth. It is connected to one terminal of a current meter (not necessarily a moving-coil meter), the other terminal of which is supplied from a voltage-regulated d.c. supply. The current meter is graduated from 'empty' at one end of the scale to 'full' at the other.

This example of the use of a float in a measurement system illustrates two points about floats in general:

(a) the float must be allowed only limited freedom of movement — in this case the arm prevents it from floating away sideways, and

(b) some kind of signal conditioning is necessary — in this case a variable resistor — to convert its displacement into some other form of signal more suitable for operating a display device.

Actually, the fuel contents measurement system in a motor car is rather a rough-and-ready device, and we learn by experience not to expect too much accuracy from it. But a float *can* be used for the accurate measurement of tank contents: Fig. 5.7 shows a float-operated system for gauging the contents of a cylindrical storage tank with a capacity of some hundreds of cubic metres of liquid.

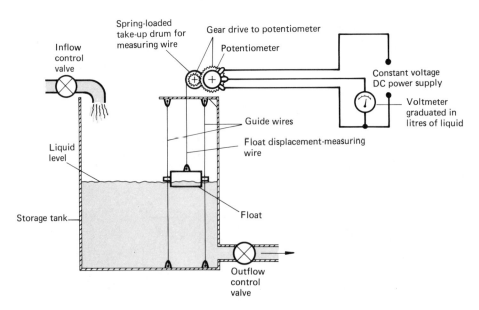

Fig. 5.7 Float-operated level-measuring system for a large storage tank

FORCE-MEASURING DEVICES

SPRING–LEVER SYSTEMS

At the beginning of Chapter 2 we considered the spring as a transducer converting force to displacement, and noted that a spring need not necessarily be in the form of a coil of wire. A beam supported at its ends and loaded in the middle, also deflects proportionately to the load applied to it, and can therefore be used as a spring. A spring of this form has a large spring rate (i.e. the ratio $\dfrac{\text{force}}{\text{displacement}}$ is large), and is usually much more compact than the corresponding coil spring would be. A spring-beam is used as the load-measuring device in a *Hounsfield Tenso-meter* tensile testing machine. It forms a very compact force-measuring transducer, which deflects only very slightly under load — so little in fact, that an amplifier must be used to amplify the deflection, to make it large enough to act also as the displayed quantity. The amplifier is a small piston-and-cylinder unit containing mercury, which, when compressed by the deflection of the beam, extrudes the mercury along a glass capillary tube. The end of the length of mercury in the capillary indicates, against a scale of force, the load being applied to the specimen. Fig. 5.8 shows the principle of this method of measuring load.

A large force can be measured on quite a small spring if it is applied to the spring through a system of levers which makes it smaller. The *Denison* tensile testing machine measures the force on the test

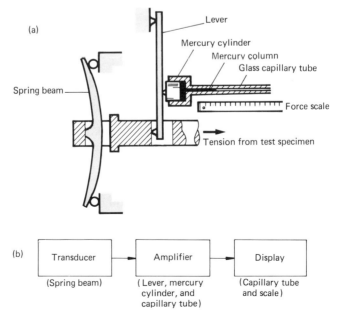

Fig. 5.8 (a) Force-measuring system of the Hounsfield Tensometer. (b) Block diagram

(a)

Lever

Mercury cylinder

Mercury column

Glass capillary tube

Spring beam

Force scale

Tension from test specimen

(b)

Transducer	Amplifier	Display
(Spring beam)	(Lever, mercury cylinder, and capillary tube)	(Capillary tube and scale)

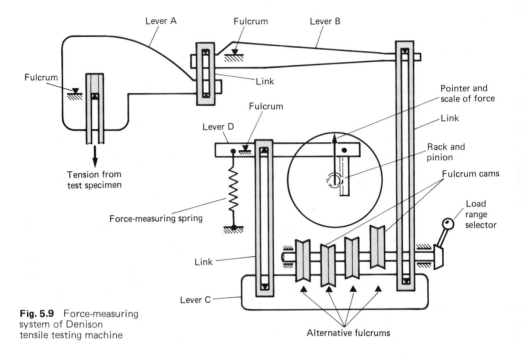

Fig. 5.9 Force-measuring system of Denison tensile testing machine

specimen in this way. Fig. 5.9 shows the arrangement of the force-measuring system, and the block diagram (Fig. 5.10) shows how it works.

The lever ratios of levers A, B and D are such that the mechanical advantage of each, as a machine, is considerably less than 1 $\left(\text{mechanical advantage} = \dfrac{\text{output force}}{\text{input force}}\right)$ so the overall mechanical advantage is very small indeed. The mechanical advantage of lever C may take one of four values, depending on which fulcrum is in use at the time.

Each of the four fulcrums in turn is brought into position by operating the load range selector lever, which rotates the shaft carrying the four cams which form the fulcrums. At each change of fulcrum, the numbers shown on the load indicating dial are automatically changed to correspond to the new sensitivity of the force measurement system.

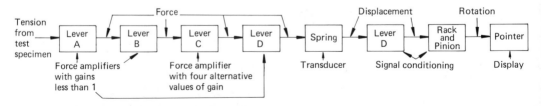

Fig. 5.10 Block diagram of the Denison force-measuring system

THE PROVING RING

If we take a steel ring and compress it or pull it along its diameter, it behaves as a spring; that is, its diameter contracts or extends in direct proportion to the force applied. The area of the radial cross-section of the ring is usually very large, and so the resulting 'spring' is very stiff. Its deflection is so small that it cannot be seen by eye, but it can be measured very accurately by means of a dial test indicator, and the change in the reading of the dial test indicator can be converted into a force measurement by means of a calibration graph, or more simply, by multiplying by the gradient of the graph, since the graph is a straight line.

The assembly of steel ring, dial gauge, and loading pads is called a *proving ring*. It is illustrated in Fig. 5.11. It can be used anywhere we need to measure very large forces, and as well as having the advantages of being self-contained and portable, it is such a simple mechanical device that there is nothing to go wrong with it, *provided that we do not exceed the maximum permissible load*. If we apply such a large load that we permanently distort the ring, the sensitivity of the proving ring will change, and its calibration will no longer be correct.

A typical application of the proving ring is in checking the calibration of the force-measuring systems of tensile testing machines.

Fig. 5.11 Proving ring

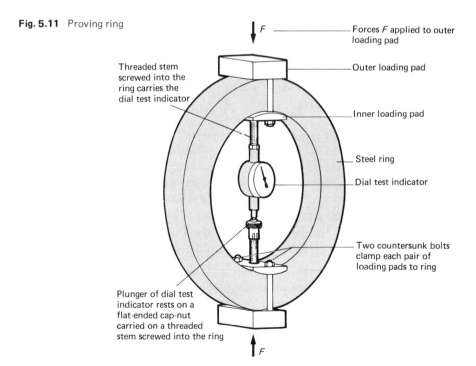

Threaded stem screwed into the ring carries the dial test indicator

Plunger of dial test indicator rests on a flat-ended cap-nut carried on a threaded stem screwed into the ring

Forces *F* applied to outer loading pad

Outer loading pad

Inner loading pad

Steel ring

Dial test indicator

Two countersunk bolts clamp each pair of loading pads to ring

STRAIN-GAUGED LOAD CELL

A load cell is a compact, self-contained, fairly rugged device for measuring a force. Load cells are often incorporated into the supports of liquid storage tanks, or weighbridges, or hoppers for feeding materials into a manufacturing process, so that they give a continuous read-out of the weight of the contents on some remote display device.

Load cells are usually of either the hydraulic type, in which the transducer converts force into pressure, or the strain-gauge type, in which force is converted into voltage. The voltage signal from a strain-gauged load cell is very suitable for use in automatic process control systems.

Fig. 5.12   Strain-gauge load cell

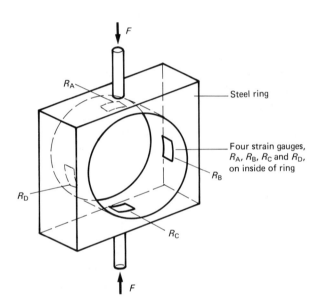

Fig. 5.12 shows a force transducer for a strain-gauged load cell. The load (F) to be measured is applied to a block of steel bored out to make it into a kind of proving ring. Strain gauges R_A, R_B, R_C and R_D are applied as shown, on the horizontal and vertical diameters, where the material is thinnest. The bending deflection at these points, due to load F, causes tensile strain in gauges R_A and R_C and compressive strain in R_B and R_D. Fig. 5.13 shows the circuit diagram and the corresponding block diagram for the complete measurement system.

In this system, all the resistances of the Wheatstone bridge circuit are active strain gauges. The result is a compact one-piece transducer of force to voltage, with four times the sensitivity of the basic single-active-gauge bridge circuit, and with complete freedom from errors due to temperature change.

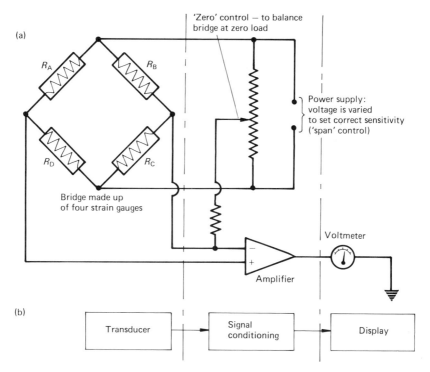

(a)

R_A

R_B

R_D

R_C

Bridge made up
of four strain gauges

'Zero' control — to balance
bridge at zero load

Power supply:
voltage is varied
to set correct sensitivity
('span' control)

Voltmeter

Amplifier

(b)

| Transducer | Signal conditioning | Display |

Fig. 5.13 (a) Circuit diagram of load cell. (b) Corresponding block diagram

Because a continuous indication of force is required, bridge balancing by the null method is *not* used — this is a direct indicating instrument.

PIEZOELECTRIC LOAD CELL

The principle of the piezoelectric transducer has already been explained in Chapter 2 (pp. 37–9), and the charge amplifier which is used with it, in Chapter 3 (pp. 79–82). Fig. 5.14 shows an example of a load cell transducer: a Kistler-Swiss type 9351 piezoelectric force link. The piezoelectric crystal is in the form of quartz discs, in contact with a miniature coaxial cable socket projecting from the side of the link. The quartz discs are preloaded in com-

Fig. 5.14 Piezoelectric
force transducer

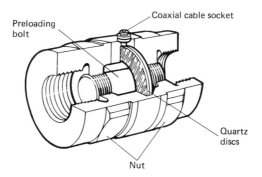

Coaxial cable socket

Preloading
bolt

Quartz
discs

Nut

pression by screwing up the nuts on the preloading bolt. This allows the force link to be used to measure tensile as well as compressive forces, because a tensile force is sensed as a reduction in the compressive force of the preload.

Piezoelectric load cells are very compact, and can be used to measure quite small forces with good accuracy. They are ideal for measuring dynamic forces, as they have very little inertia, but can measure static forces only over short time intervals.

PRESSURE-MEASURING DEVICES

THE BOURDON TUBE

The Bourdon tube pressure gauge has already been described when it was used as an example of a measurement system in Chapter 1 (pp. 5–6). Its transducer, the Bourdon tube, is a flattened tube bent into an arc of a circle, which deflects outwards in proportion to the pressure inside the tube.

The Bourdon tube pressure gauge is a rugged, foolproof, trouble-free instrument, in almost universal use. Its only disadvantage is that, being entirely mechanical, it cannot be used for remote reading (except by running a pipe to a distant pressure gauge) or for recording pressure data, and it cannot produce a signal for use in an automatic pressure control system.

However, the transducer itself, the Bourdon tube, need not be followed by mechanical signal conditioning. One method of producing an electrical output signal is to link the end of the Bourdon tube to the wiper of a potentiometer as shown in Fig. 5.15.

The sensitivity of a Bourdon tube may be increased by increasing its length. The extra length may be accommodated in a compact form by curling the tube into several turns of a spiral or a helix.

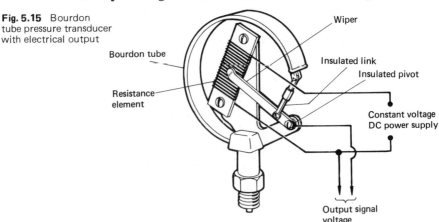

Fig. 5.15 Bourdon tube pressure transducer with electrical output

Wiper

Bourdon tube

Insulated link

Insulated pivot

Resistance element

Constant voltage DC power supply

Output signal voltage

Because a Bourdon tube has the pressure of the atmosphere acting on the outside, it measures the pressure difference between atmospheric pressure and the pressure inside the tube. That is, the measurement shown on a Bourdon tube pressure gauge is *gauge pressure*.

MANOMETERS

The basic form of manometer, the simple U-tube, is shown in Fig. 5.16. It is normally filled with either water or mercury to about half the height of the U (the 'initial level' in Fig. 5.16). If pressures p_1 and p_2 are then applied to the ends of the tube, a difference in levels, h, is produced, proportional to the pressure difference $(p_1 - p_2)$.

Fig. 5.16 U-tube manometer

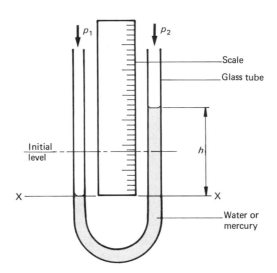

If one limb of the U is open to the atmosphere, then p_2 is exerted by atmospheric pressure, and the manometer measures the difference between p_1 and atmospheric pressure — that is, it measures the *gauge pressure* of p_1.

The pressure difference measured on the U-tube is often just stated as so many millimetres of mercury (Hg) or of water (H$_2$O), depending on which liquid was used in the manometer. If it is required in the SI units of pressure, pascals, the conversion factor can be obtained from the fact that:

standard atmospheric pressure (101.3 kPa) corresponds to 760 mm of mercury

and that:

the density of mercury is 13.6 times that of water

Alternatively, we can use the fact that at the lower level (X–X in Fig. 5.16) the pressures in the two limbs of the U are equal. We can therefore calculate the pressure difference from the equation giving the pressure at any depth in a liquid:

$$p_1 - p_2 = \rho g h$$

where g is acceleration due to gravity, $9.8\,\text{m/s}^2$, and the Greek letter ρ (rho) is the density of the liquid. (The density of water is $1000\,\text{kg/m}^3$.)

For practical purposes, the maximum pressure which can be measured on a simple U-tube manometer is about $1\frac{1}{2}$ atmospheres. At pressures greater than this, the length of tubing and the quantity of mercury required become excessive.

SELF-TEST QUESTION 10 (Solution on p. 237)

A U-tube containing mercury has one limb open to atmosphere.

(a) For a difference in levels of 28.5 mm determine:
 (i)   the gauge pressure,
 (ii) the absolute pressure,
 in SI units corresponding to this measurement.

(b) Verify your answer to (a) (i) using the alternative method given above.

(c) What would the difference in levels have been if the U-tube had contained water instead of mercury, under the same pressure?

U-tube Manometer with Liquid Above the Mercury

When a U-tube is used to measure a pressure difference in a liquid (for example the pressure difference between the upstream and throat pressure tappings on a venturi meter) it is usual to bleed the air out of the system through bleed cocks, as shown in Fig. 5.17, so that the liquid is in contact with the mercury in both limbs of the U-tube.

In this case, the pressure of h metres of mercury above the lower level (X–X) in the right-hand limb of the U-tube is partly balanced by the pressure of h metres of the liquid in the other limb.

The pressure difference is therefore calculated as:

$$p_1 - p_2 = (13.6 - d) \times 1000 g h$$

In this formula 13.6 is the relative density of mercury, d is the relative density of the liquid above the mercury ($d = 1$ for water), and $1000\,\text{kg/m}^3$ is the density of water.

SELF-TEST QUESTION 11 (Solution on p. 237)

A venturi meter is connected to a U-tube manometer containing mercury, the system being full of liquid as in Fig. 5.17. Calculate the pressure difference between the two tapping points when a difference in mercury levels of 170 mm is observed, if the liquid above the mercury is:

(a) water,

(b) kerosene (relative density 0.80).

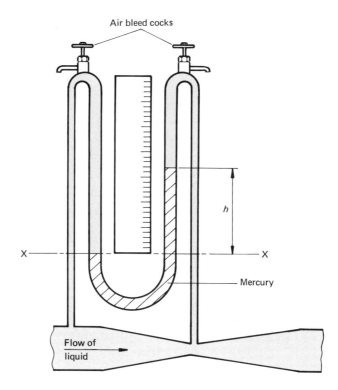

Fig. 5.17 U-tube manometer with system full of liquid

The Inclined Manometer

This kind of manometer is used to measure very small differences from atmospheric pressure, such as the 'draught' in the chimney of a boiler.

To measure such very small pressures on an ordinary U-tube, we should use water as the liquid, or, better still, a light oil, which would have a lower density than water and thus give a larger difference in levels on the U-tube. Even so, the possible error in measuring the few millimetres between the levels could become quite a large percentage error in the result.

The inclined manometer, the principle of which is shown in Fig. 5.18, reduces this error by inclining one limb at a small angle, α, to the horizontal. The effect of this is to spread out the divisions on the scale alongside it, each 'millimetre' on the inclined scale being multiplied by $\operatorname{cosec} \alpha$.

What about the other limb? The level in this is kept as nearly constant as possible by greatly enlarging the cross-section of the tube in this region, so that the displacement of liquid needed for 'full scale deflection' in the inclined limb causes negligible change of level in the enlarged limb.

Fig. 5.18 Inclined manometer

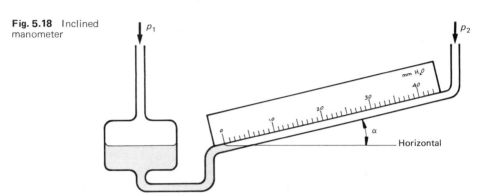

Because the reading of the manometer is very sensitive to any change in angle α, the instrument is usually mounted on levelling screws and fitted with a spirit level, so that it can be accurately set up before use.

SELF-TEST QUESTION 12 (Solution on p. 238)

An inclined manometer containing water has one limb inclined at 8° to the horizontal. The inclined limb has an inside diameter of 2.5 mm, and the enlarged portion of the other limb has an inside diameter of 38 mm. The range of the instrument is from 0 to 40 mm H₂O.

(a) Determine the length of the scale, and hence the length of one 'millimetre' scale division.

(b) Assuming that scales can be read to an accuracy of ± 0.5 mm (true length), determine the maximum percentage error when a pressure of 10 mm H₂O is measured (i) on an ordinary manometer; (ii) on this inclined manometer.

(c) Determine the change in level in the enlarged limb for 'full-scale deflection'.

The inclined limb has quite a small internal diameter, and because of this, the surface tension of the liquid in the tube will draw the level up about a centimetre above the position given by a simple pressure calculation. This effect is constant, and is taken care of by displacing the scale an appropriate amount along the tube — but the inside of the tube must be kept clean. Any contamination which alters the surface tension will cause serious errors in the readings.

A PIEZOELECTRIC PRESSURE MEASUREMENT SYSTEM

We have already considered a piezoelectric force measurement system on pp. 121–2. A piezoelectric pressure measurement system measures pressure by the force acting on the area of a small diaphragm set in the end of the pressure transducer.

As we have seen, there are much simpler and cheaper ways of measuring pressure, so why bother with a piezoelectric pressure measurement system at all? The reason is that a piezoelectric transducer:

(a) has negligible inertia so that it can accurately measure pressure variations at frequencies up to about 20 kHz;

(b) can be made very compact, so that it can be used in places where there would be no room for any other type of pressure transducer.

One application in which these qualities are essential is the cathode-ray engine indicator system. This has already been described on pp. 101–2. The pressure transducer, if it is used on a small engine, has to be very compact indeed; otherwise the engine designer may have to locate it so far from the combustion chamber that surge effects in the connecting passage distort the indicator diagram. Fig. 5.19 shows an enlarged view of a typical piezoelectric transducer.

Fig. 5.19 Piezoelectric pressure transducer for an engine indicator system

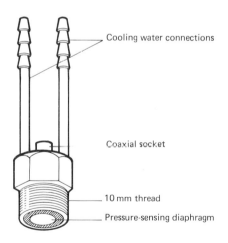

Cooling water connections

Coaxial socket

10 mm thread

Pressure-sensing diaphragm

One difficulty in designing a really compact piezoelectric pressure transducer for an internal combustion engine is the fact that the transducer must be provided with its own cooling water jacket to keep the crystal at constant temperature and thus eliminate thermoelectric drift in the signal. The water supply is usually taken from the nearest water tap, and the discharge goes to the nearest drain. Apart from eliminating drift, the cooling water supply is essential to protect the transducer from the extremely high temperatures in the combustion chamber.

CALIBRATION METHODS

FREQUENCY (ROTATIONAL SPEED) MEASUREMENT SYSTEMS

The one speed-measuring device which can be considered as a primary standard is the 60-tooth wheel, magnetic transducer and timer-counter. The timer-counter measures rotational speed by counting the number of voltage pulses it receives in exactly one second, and the only calibration necessary is a check on the accuracy with which it times its one second. This could vary a little, over the years, due to secular change (ageing) in the quartz crystal and its associated circuit.

One way of checking the accuracy of the timer-counter might be to feed it a known frequency from a signal generator and see if its measurement of frequency agreed with the setting of the frequency control on the signal generator. That would be all right if we could be sure of the accuracy of the signal generator, but the signal generator too could age.

We need a very high frequency which we can be absolutely sure of. A frequency which is stabilised for purposes such as this is the carrier wave frequency of the BBC radio transmitter at Droitwich, broadcasting on the long wave band with a wavelength of 1500 metres. The carrier wave frequency is $(300 \times 10^6)/1500 = 200$ kHz, and this is maintained constant by the BBC to within 1 part in 10^8. The transmission is receivable in any part of the British Isles, and special receivers can be obtained which will amplify the carrier wave to a strength suitable for feeding into a timer-counter.

If a timer is found to be inaccurate, the frequency of the quartz 'clock' can be adjusted by means of a trimming capacitor.

In the case of an a.c. mains-powered stroboscope, we can use the mains supply to check the accuracy of the stroboscope calibration. This is because in Britain, the electricity supply system is frequency controlled to 50 Hz, for timing purposes, with adequate accuracy for our purposes (except that we should not rely on mains timing on a very cold day, when an overload of electric heaters might temporarily slow the generators down).

The calibration of a stroboscope is checked on the little neon indicator lamp which lights up to show that the instrument is switched on. The stroboscope is set to flash at 3000 per minute (3000/60 = 50 Hz), and if the flash speed is different from the mains frequency the indicator lamp is seen to brighten and darken alternately, with a frequency (the *beat frequency*) equal to the difference between the flash frequency and mains frequency. The calibration is set right by adjusting a trimmer screw to obtain zero beat frequency.

The same procedure can be carried out at flash speeds of 12 000, 9000, 6000, 1500, etc. flashes per minute, corresponding to 4, 3,

2, $\frac{1}{2}$, etc. times mains frequency. Each adjustment of the trimmer upsets the previous adjustment, so the calibration check should be carried out only for that part of the range required for immediate use.

Stroboscopes usually have these calibration frequency points marked on the dial, together with a similar set of points based on 60 Hz, the mains frequency of the USA.

Other speed measurement systems are best calibrated using a variable-speed electric motor running at speeds which have been measured by the toothed-wheel and timer-counter system or by a stroboscope.

DISPLACEMENT MEASUREMENT SYSTEMS

Calibration checks are never carried out on dial test indicators, as the magnification factor depends on gear teeth ratios, which cannot change. If it were necessary to check the calibration, one would use tool room slip gauges, or a micrometer, to apply known displacements to the plunger of the instrument.

The linear variable differential transformer is calibrated by applying known displacements to the soft iron core and noting the corresponding output on the meter. The displacements would again be applied by slip gauges or micrometer. Care must be taken that the calibration method does not affect the magnetic coupling between the windings of the differential transformer.

A float system is easily calibrated with a rule — though it is sometimes a little difficult to decide just where the liquid level is, because of the meniscus — the curvature in the surface of the water where it clings to the rule. For precise measurements, a hook gauge is used as shown in Fig. 5.20.

Fig. 5.20 Liquid level measurement, using a hook gauge

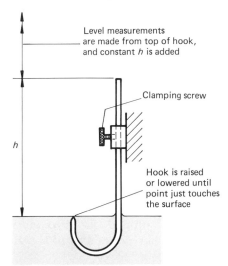

Level measurements are made from top of hook, and constant *h* is added

Clamping screw

h

Hook is raised or lowered until point just touches the surface

FORCE MEASUREMENT SYSTEMS

The most convenient way to calibrate force measurement systems is by using weights. This is not strictly a scientific way of applying a known force, because gravitational attraction varies slightly from the Poles to the Equator, and also varies with height above sea level. However, in the British Isles, g can be taken as being within the limits $9.815 \pm 0.005 \, \text{m/s}^2$, so variations in the weight of weights will be negligible for practical purposes.

Large forces can be applied and measured via a proving ring, or by using a testing machine such as the 'Denison' to apply either a tensile or compressive force.

Piezoelectric force measurement systems are a special case, because the signal from the transducer leaks away more or less quickly, depending on the quality of the components and the condition of the system. They may be calibrated with static forces applied by weights (see pp. 36–7) if the time constant of the system is long enough to permit this; if not they must be calibrated dynamically, using a vibrator to shake the transducer. A mass which is carried by the transducer, applies alternating inertia forces to it, the amplitude of which can be determined by calculation.

PRESSURE MEASUREMENT SYSTEMS

U-tube manometers are usually graduated in millimetres of mercury, or millimetres of water, or millimetres of oil of relative density d, depending on the liquid used in them. Thus they need no calibration. Where the manometer is graduated in units of pressure (newtons per square metre, or *pascals*) the calibration can be checked by calculation as in Self-Test Questions 10 and 11 on p. 124.

Bourdon-tube pressure gauges and other pressure measurement systems are normally calibrated on a dead-weight pressure-gauge tester, the principle of which is shown in Fig. 5.21.

The pressurising medium is oil, which is compressed by means of a piston operated by a screw thread. The pressure of the oil acts on the gauge to be calibrated, and also on a free-floating piston which acts as a carrier for weights. Each weight is marked with the pressure necessary to support it on the weight carrying piston. The weight carrier, also, is marked with its pressure.

To prepare the system to calibrate a pressure gauge or a pressure transducer, the screw piston is screwed in to pump oil right up to the pressure gauge seating, and then the gauge is assembled to the tester. The spring-loaded valve is then pressed down while the screw-piston is screwed in and out a few times, to ensure that all air has been pumped out through the reservoir and replaced by oil. If this is not done, all the travel of the screw-pump is used up in compressing air.

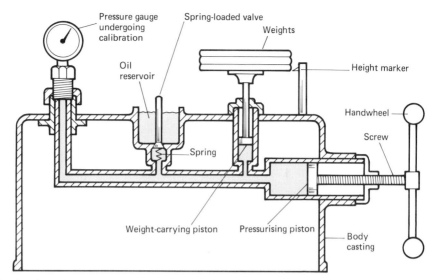

Fig. 5.21 Section through a dead-weight pressure gauge tester

To calibrate the gauge or transducer, weights are loaded on to the weight carrier, and the screw piston is screwed in until the weight-carrying piston floats up on the pressure of the oil. The gauge reading is then compared with the sum of the pressures marked on the weights and weight carrier. This procedure is repeated until enough pairs of values have been obtained to establish a calibration curve. To ensure accuracy, the stack of weights and its piston are given a twist to keep them spinning while readings are taken, so that there may be no error due to 'stiction'. And readings are only taken when the weight-carrying piston is in the middle of its travel, as shown by the underside of the stack of weights lining up with a height marker.

A piezoelectric pressure measuring system also may be calibrated in this way if its time constant is long enough to permit it. If not, the calibrating pressures must be applied more rapidly. One way in which this can be done is to use a spool valve to connect the pressure transducer to (a) a known pressure from a compressed air supply, and (b) atmospheric pressure, alternately. Rapid operation of the spool valve, at various air supply pressures, enables the relationship between gauge pressure and trace displacement on an oscilloscope screen to be determined. There is negligible loss of signal strength by charge leakage, because the signal is in the form of a square wave, alternating between levels corresponding to supply pressure and atmospheric pressure.

Where a piezoelectric pressure transducer is used as part of an engine indicator system, however, it is often used uncalibrated, because sufficient information on the effects of design changes, timing changes, etc. can be obtained by comparing the *shapes* of the pressure/volume diagrams.

SUMMARY OF FREQUENCY, DISPLACEMENT, FORCE AND PRESSURE MEASUREMENT SYSTEMS

Measuring device	Principle	Method of calibration	Remarks
(1) Eddy-current tachometer	Rotating magnet generates eddy currents in aluminium cup, causing magnetic torque on cup, proportional to rotational speed.	Calibrate against (3) or (4) on a variable-speed motor.	Used in motor-car speedometers, and hand-held tachometers.
(2) Tacho-generator	A d.c. generator with permanent-magnet field generates a voltage proportional to speed. The voltage is measured on a voltmeter graduated in rev/min.	Calibrate against (3) or (4) on a variable-speed motor.	Useful for remote-reading and automatic control systems.
(3) Stroboscope	A neon or xenon lamp is pulsed to to flash at variable frequency. When the flash rate and the frequency of the machine coincide, the machine appears stationary.	Calibrate at 3000 per min ($= 50$ Hz) or multiples or sub-multiples thereof, by obtaining zero beat frequency on a neon indicator lamp built into the instrument.	No contact with machine, therefore it absorbs no power. Possibility of error because machine also appears stationary at flash frequencies of $\frac{1}{2}, \frac{1}{3}, \frac{1}{4}$, etc. of machine speed.
(4) Pulse counter, magnetic transducer and 60-toothed wheel	60-toothed wheel makes magnetic transducer generate a.c. at frequency of $60 \times$ rev/s ($=$ rev/min). Pulse counter counts and displays this frequency.	A timer-counter's accuracy depends on its crystal-controlled oscillator for timing 1 sec. This can be checked by radio from BBC Radio 4 carrier, which has frequency of 200 kHz, accurate to better than 1 in 10^8.	The most accurate way of measuring speed but suitable only for permanent installation; also the pulse counter chops off decimal fractions of a rev/min. This could cause too large a percentage error at low speeds.

Measuring device	*Principle*	*Method of calibration*	*Remarks*
(5) Dial test indicator	A plunger operates a pointer on a scale, through a rack and pinion and a mechanical amplifier (gear train).	Positive mechanical drive between plunger and pointer so it should never need re-calibrating. Can be checked with slip gauges or micrometer.	If plunger picks up dirt, it may stick when internal return spring too weak to overcome friction. Tap the dial with a pencil to avoid this.
(6) Linear variable differential transformer (LVDT)	Displacement applied to iron core slug of a transformer with a primary and two matched secondaries. The primary is fed by an oscillator. The secondaries' outputs are anti-phase. Phase and amplitude of resultant output indicates displacement on centre-zero meter.	Apply displacement by means of a micrometer.	No contact between core and transformer, so no force exerted on object being measured. Accuracy and range can be similar to that of a micrometer. Eccentricity of core does not affect reading.
(7) Float	Used for measuring liquid level, and hence tank contents.	Displacement can be measured by means of a hook gauge and rule.	Simple device. Zero error may be caused by (i) change in density of liquid, (ii) change in weight of float due to absorption of liquid, etc.

Measuring device	Principle	Method of calibration	Remarks
(8) Spring-lever systems	Hooke's Law: displacement proportional to force, in coil springs and spring beams.	Weights or (9).	Used in spring balances and weighing machines. Hounsfield Tensometer uses spring beam, and Denison tensile testing machine uses coil spring (with lever system to scale down the load) for force measurement.
(9) Proving ring	A circular steel ring deflects proportional to compressive force across a diameter. Deflection is measured by (5) and converted to force by a calibration graph.	Weights, or a testing machine capable of applying compression, such as the Denison.	Used for measuring large compressive forces (e.g. 50 kN) with negligible deflection. Can also be used in tension. Highly accurate.
(10) Strain gauge load cell	The force to be measured is applied to a strain-gauged piece of metal.	Weights, or a compressive or tensile testing machine such as the Denison.	The bridge circuit usually has active gauges for all four resistances, bonded to a ring, or similar one-piece load carrier. Suitable for steady loads, weighbridges, etc.
(11) Piezoelectric load cell	A piezoelectric crystal acts as a capacitor which acquires charge proportional to the compressive force on it.	Weights, if they can be applied quickly enough; otherwise inertia forces applied by a mass to the load cell when it is vibrated with known frequency and amplitude.	Suitable for alternating loads or static loads of short duration.

Measuring device	Principle	Method of calibration	Remarks
(12) Bourdon tube	A metal tube with one end closed, flattened cross-section, bent into a circular arc, tends to deflect outwards to larger radius when internal pressure distends the flattened cross-section.	Dead-weight pressure tester — known pressures are applied by weights supported on a piston floating on oil, which is pressurised by a screw-operated piston.	The usual form of transducer used in pressure-gauges. Can also be used without the gauge mechanism, in automatic control systems.
(13) Vertical and inclined manometers	A glass tube bent into the shape of a U. The U is kept upright and half filled with liquid. Pressure is applied to one side; the other side is open to atmosphere. The difference in liquid levels is proportional to the pressure.	Since the pressures are usually measured in mm H_2O or mm Hg, calibration is unnecessary. If pressure is required in pascals (N/m^2) it can be calculated from $p = \rho g h$ where ρ is density of liquid and h is difference in levels.	Used with water to measure low pressures, e.g. gas supply pressures. Inclined manometer used to measure pressures close to atmospheric; e.g. chimney flue draught; air intake suctions.
(14) Piezoelectric pressure transducer	See (11)	Same as (12), if the weights can be applied quickly enough; otherwise dynamically, from known pressures alternately admitted and released by an oscillating spool-valve.	Suitable for alternating pressures, or static pressures of short duration. Used as pressure transducer in engine indicator systems — crystal housing must be water-cooled.

EXERCISES ON CHAPTER 5

1   (a) What is the name of the electromagnetic effect which occurs when there is relative motion, parallel with the surface of the material, between (i) a conductive material such as thin aluminium plate and (ii) a magnetic field which penetrates it at right angles.

(b) Explain, with the aid of a sketch, how this principle is used in a form of direct-reading tachometer.

2   Describe, with the aid of a circuit diagram, a rotational speed measurement system in which the transducer provides a voltage proportional to the angular velocity of a shaft.

3   A shaft with a radial line on its end face is rotating at 2400 rev/min. A stroboscope is aimed at the end face of the shaft and the flash frequency control is taken down through the range from 8000 to 350 flashes per minute.

(a) At what flash frequency would you expect to see:
    (i)   a triple image (radial lines at $120°$ spacing);
    (ii) a double image (radial lines at $180°$ spacing — i.e. a dia-metral line)?

(b) At what flash frequencies would you expect to see a single radial line?

4   (a) What is a stroboscope and how is it used?

(b) A stroboscope was used to measure the speed of an air turbine rotor. A stationary image was obtained at the following rates of flashes per minute:

6000,   4050,   3025,   2425,   2000,   1700

Deduce the speed of the turbine rotor assuming that there were no stationary images between these values.

5   (a) Explain how a stroboscope may be calibrated using the a.c. mains frequency as a frequency standard.

(b) A radial line on the end of a rotating shaft was illuminated by a stroboscope, and a stationary image was obtained at the following rates of flashes per minute:

5450,   2750,   1800,   1350,   1100,   900

Deduce the speed of the shaft as accurately as possible from these readings, assuming that there were no stationary images in between.

(c) For what other purpose is a stroboscope sometimes used, apart from speed or frequency measurement?

6   Draw up the truth table of a two-input 'AND' gate and use it to explain how an 'AND' gate may function as an electronic switch, to switch on or switch off a continuous stream of square-wave pulses.

7 (a) Describe, with the aid of a sketch, the toothed wheel and magnetic transducer method of indicating the speed of a rotating shaft in revolutions per minute.

(b) Under what circumstances may such a system (i) under-estimate, and (ii) overestimate the speed of the shaft?

8 (a) Show by means of a diagram how a potentiometer may be used to give a remote-reading measurement of the level of liquid in a tank. The diagram should show both the mechanical and electrical parts of the system and any source of power which may be necessary.

(b) Draw the block diagram of the system. The 'signal conditioning' part of the block diagram should include a block for each component which changes the magnitude or the nature of the signal.

(c) If the density of the liquid in the tank is altered, would this cause an error in the measurements given by the system? If so, what kind of error would it be?

(d) *Briefly* describe an alternative principle which might be used for the measurement of the contents of the tank.

9 Sketch and describe the force-measuring system of a Hounsfield tensometer.

10 Sketch and describe the force-measuring system of a 'Denison' tensile testing machine. How would you check the calibration of this system?

11 (a) Describe, with the aid of a sketch, a proving ring, and state what it is used for.

(b) What must we be careful of, when using a proving ring?

12 The calibration of a tensile testing machine was checked by using it to load a proving ring in compression. When the machine applied a load which it indicated to be 50 kN, the pointer of the proving ring dial gauge moved 60.4 divisions. Determine the percentage error in the tensile testing machine reading, if the sensitivity of the proving ring was 811.4 newtons per division.

13 (a) Make a sketch showing the arrangement of strain-gauges on a load-sensing element suitable for use in a strain-gauged load cell.

(b) Draw the electrical circuit for the complete system, identifying the strain-gauges on your sketch for part (a).

(c) Draw the block diagram for the system.

14 (a) Sketch the construction of a piezoelectric force transducer.

(b) Explain how the unit can measure tensile as well as compressive forces.

(c) What essential piece of signal conditioning equipment is required for use with a piezoelectric load cell?

15   (a) Sketch and describe the principle of operation of a Bourdon tube.

(b) When a Bourdon tube is used as a transducer, what are the basic units of (i) the input to, and (ii) the output from such a transducer?

(c) Describe, with the aid of a diagram of the apparatus used, a method of calibrating a measurement system which has a Bourdon tube as the transducer.

16   Sketch one method of converting a Bourdon tube to give an electrical output signal.

17   A U-tube manometer containing mercury has one limb open to atmosphere. The difference in levels in the two limbs is 22 mm. Convert this to:

(a) gauge pressure (Pa);

(b) absolute pressure (Pa);

(c) water gauge (mm H_2O).

Take atmospheric pressure as 101.3 kPa.

18   A U-tube manometer is used as a pressure difference measuring system, by connecting one limb to the lower pressure and the other to the higher pressure. Calculate the pressure difference corresponding to a difference in levels of 291 mm between the two limbs:

(a) if the liquid in the U is mercury, and the pressures are gas pressures;

(b) if the liquid in the U is mercury, and the remainder of the system is completely filled with water;

(c) if the liquid in the U is water, and the remainder of the system is completely filled with petrol (relative density 0.68).

19   (a) Sketch and describe an inclined manometer.

(b) What kind of measurement is this type of manometer usually used for?

(c) What precaution must be taken before taking any readings from this type of manometer?

20   An inclined manometer is to measure pressures of from 0 to 30 mm H_2O. A light oil, of relative density 0.8, will be used as the liquid, and the inclined limb will be at $10°$ to the horizontal. The bore of the inclined tube is 2 mm, and the enlarged portion of the other limb is of rectangular cross-section, with internal dimensions 20 mm $\times$ 40 mm. Calculate the actual length of the scale, between the '0 mm H_2O' and the '30 mm H_2O' divisions.

21   (a) Sketch a piezoelectric pressure transducer for an engine indicator system.

(b) What must be supplied to this transducer all the time the engine is running?

(c) Draw a diagram of a complete engine indicating system; i.e. one which displays the pressure–volume diagram of a reciprocating engine on a cathode-ray oscilloscope.

22   How would you check the calibration of:

(a) a digital timer-counter;

(b) an LVDT displacement measuring system;

(c) a tank contents measuring system for liquids;

(d) a force-measurement system which does not include a piezoelectric transducer;

(e) a piezoelectric pressure measurement system with a 'long' time constant?

23   Choose from the list at the end of this question the most suitable means of checking the calibration of:

(a) an eddy-current tachometer or a tachogenerator;

(b) a stroboscope;

(c) a linear variable differential transformer;

(d) a system for measuring the level of liquid in a tank;

(e) a tensile testing machine;

(f) a pressure gauge.

Choose from:
(i)   Venturi meter;
(ii)   proving ring;
(iii)   light-meter;
(iv)   thermocouple;
(v)   a.c. mains frequency;
(vi)   toothed wheel, magnetic transducer and counter;
(vii) micrometer;
(viii) friction brake;
(ix)   'dead-weight' tester;
(x)   hook gauge;
(xi)   ammeter;
(xii) viscometer.

Chapter 6

An Amplifier Project

In order to get some experience of electronics in general, and amplifiers in particular, students are recommended to construct and test their own amplifiers. This can be done very simply and cheaply using a type 741 integrated circuit operational amplifier (or, to use the jargon, 'a 741 i.c. op. amp.'). Fig. 6.1 shows the circuit diagram of the complete amplifier system. The clever stuff has already been done inside the integrated circuit, and all we have to do is connect up power supplies, input and feedback resistors, and a nulling potentiometer.

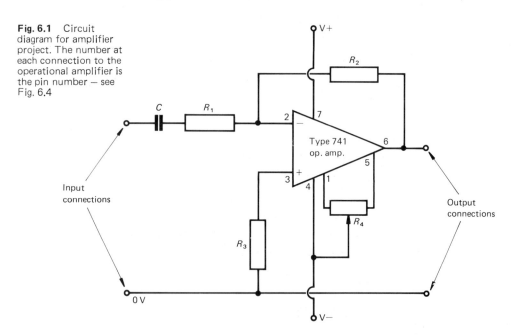

Fig. 6.1 Circuit diagram for amplifier project. The number at each connection to the operational amplifier is the pin number — see Fig. 6.4

The type 741 operational amplifier has built-in frequency compensation, and is protected from damage if the output should be accidentally short-circuited, or the input overloaded.

Without the feedback resistor, R_2, its frequency response curve is as shown by the full line in Fig. 6.2; the effect of R_2 is to reduce the gain but increase the bandwidth, as shown by the dotted line.

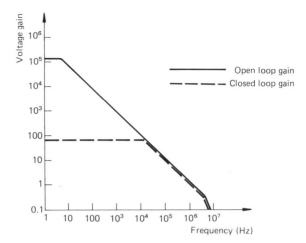

Fig. 6.2 Frequency response curve of the type 741 operational amplifier

The type 741 amplifier will work down to zero frequency; that is, it will give steady amplification of a d.c. voltage, so, strictly speaking there is no need for the coupling capacitor C. I have introduced it, however, to give a frequency response curve like that shown in Fig. 3.17 (p. 64), so that the flat part of the frequency response curve will have both lower and upper cut-off frequencies.

THE CIRCUIT

For the circuit shown in Fig. 6.1:

Resistance R_1 should be about $10\,\text{k}\Omega$.

Resistance R_2 is determined by the gain we require from the amplifier over the flat part of the frequency response curve. This is set by the ratio R_2/R_1.

Resistance R_3 should be made equal to R_2.

R_4 is a $10\,\text{k}\Omega$ miniature preset potentiometer, used for *nulling* the amplifier. It is adjusted to set the output voltage to zero, before anything is connected to the input connections.

The amplifier will work with power supplies of from $\pm 2\,\text{V}$ to $+20\,\text{V}$. A power unit supplied from a.c. mains, with connections for say, $+12\,\text{V}$, $0\,\text{V}$, and $-12\,\text{V}$ would be ideal. It must, however, have a fully smoothed and voltage-regulated output, otherwise $50\,\text{Hz}$ mains hum will mask the frequency we are looking for in the output. If a suitable power unit is not available, two PP9 batteries can be used, with the positive stud of one, and the negative end of the other, connected to $0\,\text{V}$, as shown in Fig. 6.3.

THE VARIATIONS

Two amplifiers with different characteristics should be constructed, so that the effect of connecting them in series can be investigated. The following values are suggested:

Amplifier 1

$$C = 0.47\,\mu\text{F}$$

$$R_2 = R_3 = 150\,\text{k}\Omega$$

Amplifier 2

$$C = 0.1\,\mu\text{F}$$

$$R_2 = R_3 = 220\,\text{k}\Omega$$

THE SHOPPING LIST

The complete list of components for the two amplifiers is given below. To assist people who might be bewildered about what to order, the RS Components Ltd. stock number is given against each item, though obviously the components can be obtained from any other suitable supplier. The total cost per two amplifiers is difficult to predict in terms of money, because of possible inflation. In terms of first-class postage stamps for letters (an inflation-proof currency?), and excluding VAT, the two amplifiers will cost a total equivalent to about 50 such stamps at present day prices.

RESISTORS

Carbon composition or metal film; any wattage will do, provided they do not take up too much room.

2 off: 10 kΩ (Colour coded brown/black/orange) for R_1 (RS Stock No. 142–485).

2 off: 150 kΩ (Colour coded brown/green/yellow) for R_2 and R_3 of amplifier 1 (RS Stock No. 142–637).

2 off: 220 kΩ (Colour coded red/red/yellow) for R_2 and R_3 of amplifier 2 (RS Stock No. 142–659).

2 off: 10 kΩ carbon track preset miniature horizontal potentiometer, for R_4 (RS Stock No. 185–000).

CAPACITORS

(Any sort will do, provided they are *not* electrolytic, and not too big. I suggest metallised polyester film capacitors. These are colour coded in bands with the colours of the first three bands as shown below.)

Capacitor for amplifier 1: 0.47 μF (banded yellow/violet/yellow) (RS Stock No. 113–926).

Capacitor for amplifier 2: 0.1 μF (banded brown/black/yellow) (RS Stock No. 113–904).

AMPLIFIER

2 off: type 741 operational amplifier (RS Stock No. 305–311).

HARDWARE

2 off: standard 8-way dual-in-line socket (RS Stock No. 401–683).

4 off: matrix board (SRBP board); 0.1″ hole spacing; 104 × 65 × 1.6 mm, or any similar standard size. (One as insulating base, one as circuit board, for each amplifier.) (RS Stock No. 433–602.)

Terminal pins, press fit, single sided (RS Stock No. 433–624).

Tinned copper wire 20 s.w.g. (RS Stock No. 355–063).

Flexible connecting wire (e.g. 16/0.2 mm stranded copper wire, PVC sheathed.) (RS Stock No. 356–549).

4 pairs PP9 battery clips* (RS Stock No. 488–012).

2 off: switch* — at least 3-pole (RS Stock No. 338–636).

BUILDING THE AMPLIFIERS

The amplifiers are constructed on matrix boards (Veroboard) with 0.1″ hole spacing as shown in Fig. 6.3.

Wiring which is above the board is shown thus —————; wiring on the underside of the board is shown thus ------------------- ; the holes through which the wires pass from one side to the other being indicated thus ————o---------- .

The electrical connections to the type 741 integrated circuit are actually made to an 8-pin dual-in-line (or d.i.l.) socket, and the integrated circuit is only inserted into the socket after all soldering is completed. This is to avoid damaging the integrated circuit with the heat from the soldering iron, and to allow easy substitution of a new integrated circuit if it is suspected that the existing one is faulty.

At points P, the wiring is soldered to single-sided terminal pins (Veropins) specially made for the matrix board, which are pressed into the matrix holes from the underside.

The resistors and the capacitor are secured by their own wires. The

*These components are only necessary if batteries are to be used as the power supply.

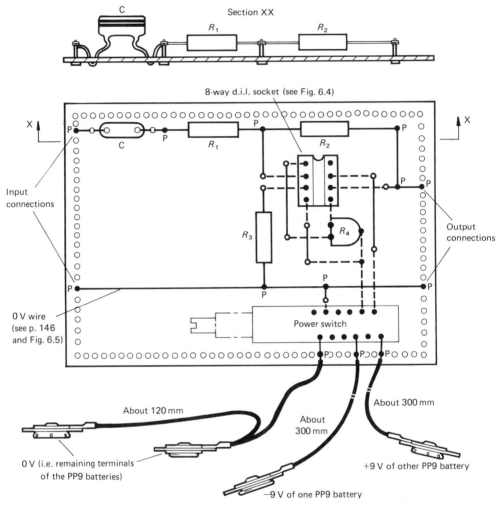

Fig. 6.3   Layout of amplifiers

remaining wiring is done with 20 s.w.g. tinned copper wire, which is rigid, but thin enough to be bent easily.

The connections are made by soldering. This should be done using a miniature soldering iron, and multicore solder, which contains its own flux. Wherever possible, a mechanical connection should be made first, by hooking the wire round the terminal pin or hooking wires together, before solder is run into the joint to make the electrical connection. This will be impossible, however, in the case of the amplifier socket, the potentiometer, and the switch. To make the connections to these, the ends of the tags projecting through the holes in the board should be bent over as close as possible to the underside of the board, and 'tinned' with solder. The wires to be connected to them should also be tinned with solder; it is then an easy matter to 'melt' each wire on to its tag to make the joint. The connections from the switch to the three terminal pins at the edge of the board are made in a similar way: the undersides of the terminal pins are tinned, the corresponding three tags on the switch are tinned, then the tags are bent over on to the pins and 'melted' on to them.

After construction is complete, the undersides of the amplifiers should be covered with another sheet of matrix board, wired or bolted to the upper board, to insulate it from accidental short-circuits from discarded ends of wire, solder, etc., which might be lying about the bench.

CHECK YOUR WORK

Before connecting up the power supplies and plugging the amplifier into its socket, we must check everything thoroughly. The finished circuit must first be checked against Figs. 6.1 and 6.3 to make sure that we have not made any mistakes in our wiring. Then using a multimeter set to the lowest resistance range we must check the quality of our soldering by checking the resistance across each soldered joint ($0\,\Omega$ should be obtained each time). The soldering to the tags of the amplifier socket can be checked by pushing one strand from the flexible connecting wire into each of the eight small sockets in turn, to make contact with the tag, then measuring the resistance to the wire soldered to the tag — it must be $0\,\Omega$ again. (If your soldering tends to be messy, it might be as well to check, at the same time, that the resistance to the wire on either side of the tag you are checking is infinity!) Finally, check continuities through the switch.

NULLING THE AMPLIFIERS

When the checking is finished, the 741 amplifiers can be plugged into their sockets on each amplifier board. Take care to get them the right way round — Fig. 6.4 shows that there is a notch at the end, between pins 1 and 8 and there should also be a spot or dimple above pin 1. The notched end of the amplifier should face resistance R_2 (see Fig. 6.3).

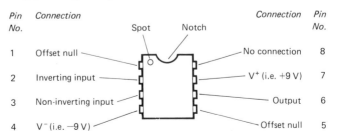

Fig. 6.4 Enlarged view on top of the type 741 operational amplifier, showing pin connections

Pin No.	Connection			Connection	Pin No.
		Spot	Notch		
1	Offset null			No connection	8
2	Inverting input			V^+ (i.e. +9 V)	7
3	Non-inverting input			Output	6
4	V^- (i.e. −9 V)			Offset null	5

Each amplifier must now be 'nulled'. This means adjusting the setting of potentiometer R_4 so that no output voltage is produced when no voltage difference exists at the input. The procedure is as follows:

(a) Ensure that there is nothing connected to the pins marked 'input connections' in Fig. 6.3.

(b) Set the wiper of the nulling potentiometer, R_4, to about the mid-point of its travel.

(c) Connect up the batteries or the power supply and switch on.

(d) Connect a high-resistance multimeter to the pins marked 'output connections' in Fig. 6.3. Set it to its lowest d.c. voltage range and adjust R_4 to bring the voltmeter reading as near zero as you can. If the voltmeter reading is already zero, try changing over the voltmeter connections, in case the pointer is trying to go in the wrong direction.

Unless you have a very sensitive meter, you will probably find that the setting of R_4 is not at all critical — in fact the main purpose of the nulling procedure is really to make sure that there is no excessively large output voltage that cannot be nulled out. If there is, go back and check everything through again, looking for a fault in your wiring. As a last resort, replace the 741 integrated circuit with another one.

DETERMINING THE FREQUENCY RESPONSE

A signal generator (or function generator) and a double-beam cathode ray oscilloscope are needed for this. These are connected to the amplifier as shown in Fig. 6.5. Ordinary flexible connecting wire, soldered to the input and output connection pins of the amplifier board, will probably be all that is needed to connect the amplifier to signal generator and oscilloscope. Take care, however, to make these connecting wires as short as possible (within reason). and keep the output wires well away from the input, to avoid distortion of the waveform and (possibly) oscillation, which could occur if the amplifier picked up and reamplified a significant proportion of its output. If this appears to be happening, coaxial cable (screened cable)* may be used instead, for all connecting leads. The braided sleeve of this should be connected to the earth (black) terminals on signal generator and oscilloscope, and soldered to the zero volt wire on the amplifier board. The centre wire of the coaxial cable is then used to carry the signal from the signal generator to the remaining input terminal on the amplifier board, and from the remaining output terminal to the Y_1 input of the oscilloscope.

Another length of coaxial cable could be similarly used to carry the signal direct from the signal generator to the Y_2 input of the oscilloscope, for comparison with the output of the amplifier.

*RS Stock No. 367–280 is suitable, but any other coaxial cable would do equally well.

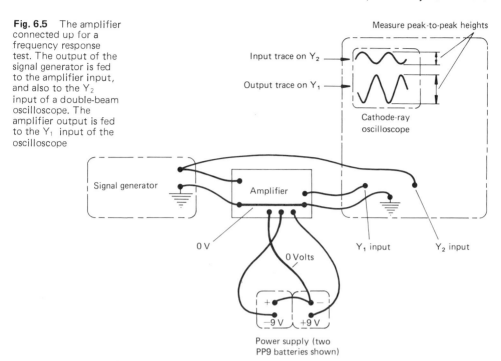

Fig. 6.5 The amplifier connected up for a frequency response test. The output of the signal generator is fed to the amplifier input, and also to the Y$_2$ input of a double-beam oscilloscope. The amplifier output is fed to the Y$_1$ input of the oscilloscope

It is important that the black (or earthed) terminal of the signal generator is connected, through the zero volt wire, to the black (or earthed) terminal on the oscilloscope, because these terminals are often connected to the chassis of the instrument, which in turn is connected to the earth pin of the mains a.c. power socket. In that case, changing over the connections at signal generator or oscilloscope would short-circuit the input or output of the amplifier, causing a mystifying loss of signal.

If either or both instruments have coaxial sockets, the rule is to connect outer conductor to outer conductor, as the outer conductor of coaxial cable is always the 'earthy' one.

When all the connecting up has been completed, the oscilloscope should be given a timebase setting of about 1 ms/cm, with triggering from the Y$_2$ input, and switched on so that the resulting horizontal-line traces can be located and focused. The Y$_2$ input should be set to maximum sensitivity, and the Y$_1$ to about 5 V/cm. Both Y$_1$ and Y$_2$ should be switched to 'd.c.' because 'a.c.' introduces a capacitor which might affect readings at low frequencies.

The signal generator should then be set to a frequency of about 1 kHz and an output voltage of zero, and switched on. If its output voltage is now gradually increased, the upper sine wave shown in Fig. 6.5 should be obtained. The signal generator output should be kept as low as possible, consistent with getting a measurable sine wave; otherwise, when the amplifier is switched on, its output may be 'clipped' or otherwise distorted because it is overloaded.

If the amplifier is now switched on, the lower sine wave shown in Fig. 6.5 should be obtained. The two sine waves will be 180° out of phase because the signal is applied to the *inverting* input of the operational amplifier.

The frequency response test determines the gain of an amplifier at various frequencies covering the useful frequency range of the amplifier. The gain is calculated in decibels, using the formula on p. 62. The voltage ratio in this formula is, strictly speaking, a ratio of amplitudes, but it is much easier to measure peak-to-peak values as shown in Fig. 6.5, and of course these are in the same ratio as the amplitudes. Measuring peak-to-peak values, we need not bother too much about the triggering of the oscilloscope or about its timebase setting — if all we get on the screen is a blurred band because the timebase is too slow and/or untriggered, we can still measure the height of the band, and hence the peak-to-peak value. Of course, it is better to get clear sine waves if we can, to be sure that no distortion or clipping is occurring.

To make our measurements as accurate as possible, we should make full use of the Y-shift controls, shifting each trace in turn off the screen, so that the other can be expanded, by means of the V/cm switch, to fill the full height of the screen as far as possible.

The table opposite shows a suitable layout for the table of results from a frequency response test.

The following notes should make the table clear:

The frequencies shown in Column 1 are the most suitable for covering the required frequency range with the minimum number of measurements, and give roughly equal spacing on a logarithmic scale of frequency;

The row of values for 600 Hz is shown filled in with typical experimental results, to show how the gain values are calculated;

Columns 2, 3, 5 and 6 are filled up during the test, the values in the remaining columns being calculated from them;

If the signal generator is a reasonably good one, columns 2 and 4 will show little or no change over the full frequency range of the test, the value in column 3 being the same for all frequencies.

At the top of the table are spaces for entering an amplifier identification number and the values of R_1 and R_2, from which the theoretical voltage gain of the amplifier can be calculated. The amplifier identification should also be marked on the matrix board of the amplifier, so that there may be no doubt as to which table of results relates to which amplifier. The values of R_1 and R_2 should be their actual values as measured by an ohmmeter while the integrated circuit is removed from its socket, as there may be up to 10% variation from the nominal values.

Amplifier No. .... R_1 = ... kΩ R_2 = ... kΩ   ∴ max voltage gain expected = R_2/R_1 = ...

(1)	(2)	(3)	(4)	(5)	(6)	(7)	(8)	(9)
	Amplifier input (Y_2) values			Amplifier output (Y_1) values				
Frequency (Hz)	Peak-to-peak height (cm)	Sensitivity (V/cm)	Peak-to-peak voltage (V)	Peak-to-peak height (cm)	Sensitivity (V/cm)	Peak-to-peak voltage (V)	Voltage gain $\left(\dfrac{V_{out}}{V_{in}}\right)$	Gain (dB) $\left[= 20\log_{10}\left(\dfrac{V_{out}}{V_{in}}\right)\right]$
10								
20								
40								
60								
100								
200								
400								
600	6.25	0.1	0.625	5.9	2.0	11.8	18.88	25.5
1000	*mol*							
2000								
4000								
6000								
10 000								
20 000								
40 000								
60 000								
100 000								

The right-hand column (9) of the table is used to plot the frequency response curve of the amplifier. This should be plotted on 4-cycle log-linear graph paper as shown in Fig. 6.6. If log-linear paper is not available, the 'log' function on a scientific calculator can be used to obtain the logs-to-the-base-10 of the frequencies, and these can be marked off along the horizontal axis on ordinary graph paper, and labelled with the corresponding frequencies.

Fig. 6.6   Typical curves obtainable from the frequency response tests. The actual curves will depend on *actual* values of *C*, *R*$_1$ and *R*$_2$

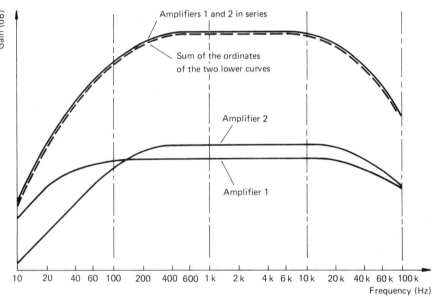

If two amplifiers with two distinct values of gain are constructed, as has been suggested, the two lower curves of Fig. 6.6 will be obtained. Now comes a slightly more difficult test. We put the two amplifiers in series, so that the output of the first is the input to the second, and do a frequency response test on this combination. We should obtain the upper full curve of Fig. 6.6 and, if everything goes right, this should coincide with the dotted curve, obtained by adding corresponding ordinates of the two lower curves. This verifies that the overall gain of two components in series is the sum of their decibel gains (which corresponds to multiplying their voltage gains).

It is a more difficult test because the overall voltage gain of the two amplifiers in series could be up to 500 V, and instability is more likely to occur at this level of gain, especially at the higher frequencies. In real life, we should overcome this by arranging a single feedback link from the output of the second amplifier to the input of the first, but this would nullify the comparison we are trying to make. So we must get over it by cutting down the input to the first amplifier to an absolute minimum. An easy way of doing this is to apply the signal generator's output to the outer terminals of, say, a 1 kΩ preset potentiometer (A and B in Fig. 3.26 on p. 76) and take the input to the first amplifier from one outer terminal and the centre terminal, (A and C in Fig. 3.26). If the potentiometer has been pre-set, using a battery and a voltmeter, to make V_{AC} equal to one tenth of V_{AB} then we shall be applying one tenth of the signal generator's output to the first amplifier's input.

Of course when we come to calculate gain, we shall have to remember that voltages shown by the Y_2 trace of the oscilloscope in this case are ten times the actual input voltage to the first amplifier.

PART TWO

Control

Chapter 7

Principles of Automatic Control

CONTROL STRATEGY

People spend their lives trying to control things — with varying degrees of success.

The engineer, too, is concerned with control. He (or she) may have to control a machine, or a process, or a system; or may have to design one so that a driver or operator can control it. Either way, 'control' to an engineer means something much more precise than it does to most other people.

Ideally, it means *exact* control; the output of the machine, process or system instantly following the 'demand' setting of the input, without deviating from it in the slightest.

In practice, such instant, absolute precision is unattainable. The engineer controls the system sufficiently for the deviation of the output from the demanded value to be within acceptable limits. The smaller the allowable variation in the output, the more complicated (and therefore more costly) the control arrangements will have to be.

OPEN-LOOP CONTROL

This is the simplest form of control. The principle is that *at some point in time, we control the output of a system as accurately as we can — and then leave it to work on its own*. A bullet fired at a target, for instance, is under open-loop control from the instant it leaves the gun. If we have made an error in aiming it, or if there is a 'disturbance' — say, a sudden gust of wind while the bullet is on its way — there is no way that the output will be altered to correct for it.

It is a form of control which works tolerably well in some situations — where we cannot arrange or cannot afford a more sophisticated form of control. A lathe, for instance, may be set up for a particular speed, feed and depth of cut, and then be set in motion and left to complete the cut unattended. And provided there are no 'disturbances', the result will be satisfactory.

In many situations, though, open-loop control would be disastrous.

To illustrate the principles of automatic control, let us start with a simple open-loop control system and 'build it up' into automatic control. To do this, we shall have to represent the system by a block diagram. All automatic control systems have to be simplified into block diagram form, otherwise we should never be able to comprehend them.

I will take as an example a hydraulic power unit, consisting of a petrol engine driving a hydraulic pump. The block diagram of the unit under open loop control is shown in Fig. 7.1.

Fig. 7.1 Block diagram of hydraulic power unit under open loop control

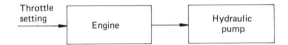

If the system were truly allowed to run unattended under this kind of control, external disturbances such as a change in the demand for the output of the hydraulic pump could cause the engine either to 'race' to destruction, or to 'stall'.

In practice, even in the case of the lathe used in the earlier example, somebody would (or should!) be keeping an eye on it. Certainly, in the case of the hydraulic power unit, if the speed varied too drastically, a human-being would have to be at hand to 'close the loop', as shown in Fig. 7.2.

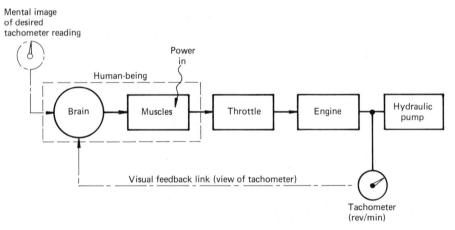

Fig. 7.2 Block diagram of hydraulic power unit under continuous manual control

The human-being would act as a *controller*. His (or her) brain would be the *error detector*, comparing the actual reading of the tachometer with his mental image of what the reading ought to be. His brain would actuate his muscles to adjust the throttle so as to make the actual speed coincide with the intended speed. That is, in terms of control theory, the *error* or deviation signal he would send to his muscles would be a function of the difference between the *demand* or *reference* signal and the signal fed back visually from the *measured* or *output value* shown on the tachometer.

His brain alone could not adjust the throttle setting; its error signal needs *power amplification*, which is provided by his muscles.

Everybody, more or less frequently, finds himself (or herself) acting as the controlling component of this kind of system — when driving a car or riding a motor cycle, for instance. It is a manual-control closed-loop system, but it is *not* true automatic control. For automatic control, we must replace the human component by something more consistent and more reliable.

CLOSED-LOOP CONTROL

For true closed-loop control, there must be:

(a) *negative feedback*, in which a feedback signal, which is a function of the output value to be controlled, is subtracted from the reference signal;

(b) *error actuation*, in which the difference between the reference signal and the feedback signal (the *error*) is used to adjust the output value to its required level;

(c) *no human intervention* once the reference signal value has been set.

The block diagram (Fig. 7.3) shows how the original open-loop hydraulic power unit might be converted to closed-loop control.

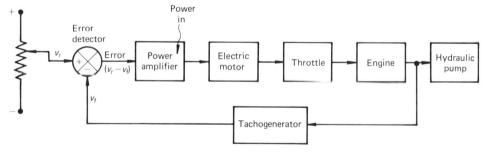

Fig. 7.3   Block diagram of hydraulic power unit under closed-loop control

The reference signal might be a voltage, v_r, set by the wiper of a potentiometer. A tachogenerator driven from the engine could act as a transducer to give a feedback voltage v_f proportional to engine speed. The error detector could be an operational amplifier as in Fig. 7.4, with the non-inverting input earthed through R_3, and v_r and v_f applied to the inverting input through resistances R_1 and R_2 respectively. Subtraction of v_f from v_r is obtained by connecting the tachogenerator so that its output voltage v_f is negative, and more or less cancels out the reference voltage v_r, from the wiper of the potentiometer, which is positive. The output of the operational amplifier would go to the power amplifier, and the output of the power amplifier would be strong enough to power an electric motor which would move the throttle, through some kind of gearing.

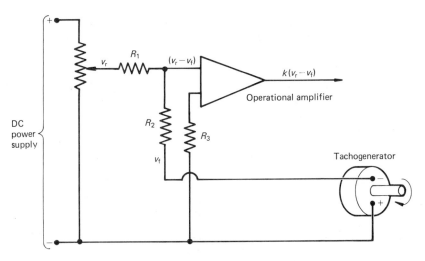

Fig. 7.4 Using an operational amplifier as error detector for electrical signals; $R_1 = R_2 = R_3$; the gain of the amplifier is k

If the load on the engine increases, causing its speed to fall, the feedback signal, v_f, becomes less than v_r, so the error signal, $v_r - v_f$, changes from zero to a positive value.

This causes the power amplifier to pass current through the electric motor so that it opens the throttle to increase engine speed. The process continues until v_f becomes equal to v_r again, with the engine speed back to its original setting.

If a load change causes the engine speed to rise, $v_r - v_f$ becomes negative. This causes the power amplifier to pass current through the electric motor in the opposite direction, causing it to close the throttle until v_f equals v_r again.

INSTABILITY

The system I have just described, the system represented by the block diagram of Fig. 7.3, probably seems so straightforward that it could not fail to work correctly. And yet if we actually built it, using any components that came to hand, while we might be lucky and find that it was *stable*, with a response of the form shown in Fig. 7.5(a) or (b), it could just as easily be *unstable* and cause the speed to oscillate wildly, as shown in Fig. 7.5(c).

The designer of an automatic control system must be able to use mathematical and graphical techniques to find out, *before* his proposed system is built, whether it will be unstable or not, and if so, to design the instability out of the system. The necessary techniques are beyond the scope of this book, but we should be able to see *why* a control system may become unstable.

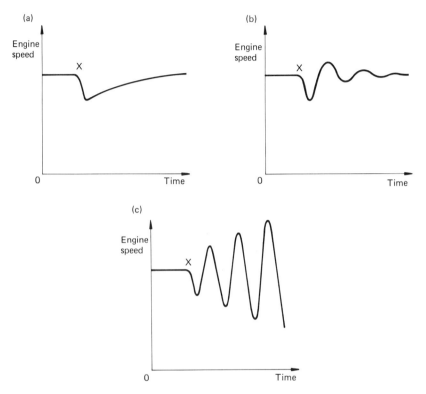

Fig. 7.5 Possible responses of the closed-loop control system of Fig. 7.3 when a sudden increase of load is applied to the engine at the instant marked X: (a) heavily damped, stable; (b) lightly damped, stable; (c) unstable

The cause of the instability is the fact that, in a closed-loop system, the output of each block in the block diagram forms the input to the next block, and when controlling action is required, the necessary changes take time to work their way through each block from input to output.

Suppose we have a sine wave input applied to one of the blocks, as shown in Fig. 7.6(a). The output will have the same frequency as the input, but due to the time delay there will be a phase lag, an angle ϕ, calculated as shown in Fig. 7.6(b). Also, the block will have a gain at that frequency, given by the ratio of the output amplitude to the input amplitude.

The phase lag and gain of a block depend on the frequency applied to it.

Now the phase lags of all the blocks in the closed loop add together, while the gains of all the blocks are multiplied together. If, at any frequency, the total phase lag in the loop is 180° while at the same time the product of all the gains (the *loop gain*) is 1.0 or greater, then we shall have the situation shown in Fig. 7.7(c).

Fig. 7.6 (a) Sine-wave input applied to a block, and the corresponding output. (b) Time-graphs superimposed, showing the calculation of phase lag and gain for a particular frequency

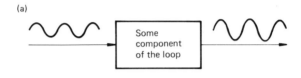

(a)

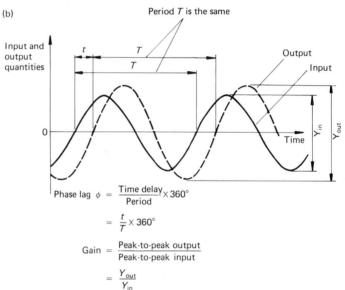

(b)

Period *T* is the same

Input and output quantities

Output

Input

$$\text{Phase lag } \phi = \frac{\text{Time delay}}{\text{Period}} \times 360°$$

$$= \frac{t}{T} \times 360°$$

$$\text{Gain} = \frac{\text{Peak-to-peak output}}{\text{Peak-to-peak input}}$$

$$= \frac{Y_{out}}{Y_{in}}$$

Fig. 7.7(a) shows a sinusoidal disturbance of the output of the system (i.e. a ripple in the engine speed/time graph, in this case). Fig. 7.7(b) shows the correction which would be applied to the system if we had negative feedback with zero phase lag and with a loop gain of 1.0. It is of exactly opposite phase to the original disturbance. Of course, to be able to see it like that, we would have to open (i.e. break) the loop — otherwise the correction would kill the original disturbance instantly. Fig. 7.7(c) shows the same correction as Fig. 7.7(b) but with a phase lag of 180°. This is now exactly in step with the original disturbance (a) and would keep the ripple going indefinitely, with constant amplitude. If the loop gain was greater than 1.0, the amplitude of the ripple would increase with time, while a loop gain of less than 1.0 would cause it to die away. So at 180° phase lag, the loop gain of 1.0 is the boundary between stability and instability. Of course, if the loop gain was only *just* under 1.0, the system would probably appear to be pretty unstable still, and control engineers usually design to have a considerable *gain margin*; i.e. design the system to have a loop gain of considerably less than 1.0, at the frequency at which the phase lag is 180°.

Fig. 7.7 Effects of feedback

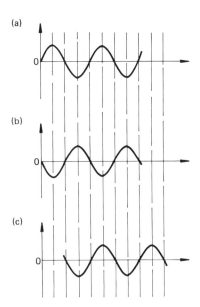

(a)

(b)

(c)

What if we just take care not to excite the system at the particular frequency at which it becomes unstable — will it then behave itself? The answer, I am afraid, is that it is impossible *not* to excite it at that frequency. Random electrical 'noise', or even a pure step input, can be shown to be made up of an infinite number of component sine waves, one of which is bound to be at the frequency we are trying to avoid.

The only way we can eliminate the instability is either to reduce the loop gain — this will unfortunately make the response of the system more sluggish — or else to reduce the phase lag around the loop.

Referring back to Fig. 7.7(c) again, notice that the effect of correcting action delayed by a phase lag of 180°, while the gain is 1.0 or more, is the same as if we had used *positive feedback* (i.e. *adding* the signal in the error detector) in a system with no phase lag.

DEFINITIONS

(Reference: *BS 1523: Glossary of terms used in automatic controlling and regulating systems*).

Demand or **reference signal:**   the command signal input to an automatic control system. It goes to the error detector and sets the value which the *controlled condition* (the output of the control system) ought to have. In Fig. 7.3, v_r is the demand signal; it sets the required value of engine speed.

Negative feedback: the transmission of a signal from a later to an earlier stage, to be *subtracted* from the reference signal in the error detector. In Fig. 7.3, v_f is the feedback signal.

Positive feedback: if the feedback signal were *added* to the reference signal we should have positive feedback. The effect of this would be to drive the controlled condition rapidly to either its upper or its lower limit. There it would remain unless the reference signal could be set beyond that extreme — the controlled condition would then swing over to the other extreme. The principle is used in some electronic circuits called *flip-flops*.

As explained on the previous page, at some frequency a negative feedback system may have a total phase lag, around the loop, of 180°. This makes it, in effect, a positive feedback system at that frequency, and if the loop gain is 1.0 or more, the system becomes unstable.

Error or **deviation signal:**   the output of the error detector; the result of subtracting the feedback signal from the demand signal. ($v_r - v_f$ in Fig. 7.3.)

Power amplification: increasing the power of a signal, so that it is capable of doing some physical work. See the notes on voltage amplifiers and power amplifiers on pp. 55-6.

A **process control system** controls some physical quantity or condition of a continuous manufacturing process (e.g. control of flow rate in a pipeline of a chemical plant).

A **servo control system** or **servo-mechanism** controls a mechanical output such as a linear movement or a rotation. It includes a power amplifier operating a **servo-motor** (which could be an electric motor or a hydraulic jack). We have already met an example in the pen carriage servos of the XY plotter (Fig. 4.6, p. 98).

A **continuous control system** is one in which the output changes smoothly and continuously, without any steps, as the reference signal is varied.

An **on–off control** system is a simpler, cruder alternative to the continuous control system. It is a system in which the power input can only be either 'full on' or 'switched off', changing from one state to the other as the error signal changes from positive to negative.

Sequence control or **programmed control** occurs when a component provides a sequence of predetermined values of reference signal to a control system, as a function either of time or of some other variable. Modern domestic washing machines, for example, operate under programmed control.

EXERCISES ON CHAPTER 7

1 (a) Give an example, with block diagram, of an open-loop control system.

(b) What are the limitations of open-loop control systems, and what facts justify the continued use of these systems?

2 (a) State three essential features of true closed-loop control.

(b) Give an example of a closed-loop control system, and draw its block diagram.

3 What is meant by instability in closed-loop control systems? Illustrate your answer with sketched graphs showing three possible responses to a step input.

4 (a) What is the cause of instability in a closed-loop system?

(b) What are the conditions, in terms of open-loop gain and phase lag, which define the boundary between stability and instability?

(c) Show, by means of sketched graphs, that negative feedback with $180°$ phase lag around the open loop is equivalent to positive feedback with zero phase lag.

5 In Figs. E7.1 and E7.2 the plain curve shows the input to, and the dotted curve the output from, components forming part of a closed-loop control system. Calculate:

(a) the frequency of the waveform;

(b) the gain of the component at that frequency, in basic SI units;

(c) the phase lag introduced by the component at that frequency; for (i) Fig. E7.1 and (ii) Fig. E7.2.

Fig. E7.1

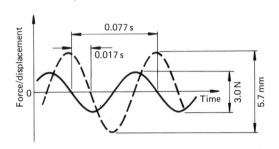

Fig. E7.2

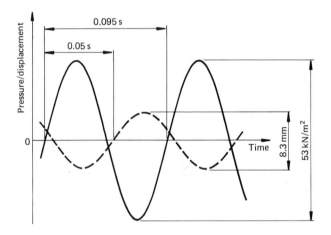

6   In a test to determine whether a closed-loop control system would be unstable, the feedback line was broken at the entry to the error detector, and the system was fed with a sinusoidal input of gradually increasing frequency. At a particular frequency, curves A and B in Fig. E7.3 were obtained from the input and the feedback line, respectively.

State whether the system is stable or unstable, explaining how you arrived at your conclusion.

7   Explain the meaning of the following terms, as used in control engineering:

(a) reference signal;

(b) error;

(c) power amplification.

Fig. E7.3

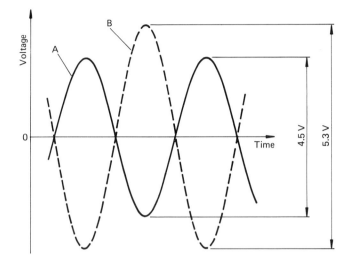

8   (a) What is meant by 'feedback'?

(b) Explain the difference between positive feedback and negative feedback. How do they differ in their effects on an automatic control system?

(c) Is it possible for negative feedback to turn into positive feedback? If so, explain how this could come about.

9   Explain how process control systems differ from servo control systems.

10 Explain the differences between (a) continuous control, (b) 'on-off' control, and (c) sequence control systems.

Chapter 8

Some Practical Control Systems

A BLOCK DIAGRAM FOR ALL SYSTEMS

In Fig. 7.3 on p. 157 we considered a possible automatic control system for regulating the speed of an engine. In practice, as we shall see, this is usually done much more simply, using a mechanical governor, but the block diagram of that system also would have the same basic form. In fact, the block diagrams of all automatic control systems using negative feedback have the same basic form, which is shown in Fig. 8.1.

Fig. 8.1 Basic block diagram of all automatic control systems using negative feedback

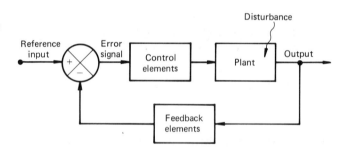

Each of the blocks in the basic diagram, above, can be expanded into component blocks to suit the particular system we have to represent. Thus for the system modelled by Fig. 7.3:

the *control elements* block expands into the three blocks: 'power amplifier', 'electric motor' and 'throttle';

Feedback elements is obviously 'tachogenerator';

Plant is obviously 'engine';

Disturbance is whatever would cause the output to change, relative to the input, if the system were under open-loop control. Thus in the system modelled by Fig. 7.3, *disturbance* is a change in load on the engine, or a change in the power output of the engine at a given throttle setting, or both.

We are now going to use this method to work out the block diagrams of some real-life automatic control systems. The first two consist of such everyday pieces of hardware that it is perhaps a little surprising to have to think of them as negative feedback systems, but that is what they are. The other examples illustrate standard engineering practice in automatic control.

AN ELECTRIC FIRE CONTROLLED BY A THERMOSTAT

The essential features of the system are shown in Fig. 8.2. The bimetal strip bends outwards as it warms up and inwards as it cools down, due to the different coefficients of expansion of the two metals which make up the strip. Setting the control knob to the temperature required rotates the cam, compressing the spring and pivoting the bimetal strip so that the electrical contacts are closed. While they are closed, they allow mains a.c. to pass through the resistance element of the electric fire, which converts the electrical energy into heat energy.

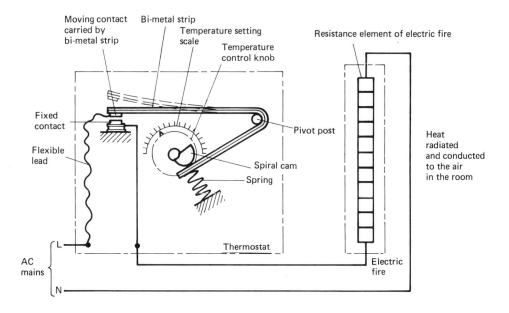

Fig. 8.2 Diagram of an electric fire controlled by a thermostat

As the air in the room warms up, the bimetal strip warms up as well and starts to curl outwards reducing the force pressing the contacts together. Eventually, the contacts flip open (by overcoming the pull of a small magnet — not shown), switching off the electric fire, and the system starts to cool.

SELF-TEST QUESTION 13 (Solution on p. 239)

(a) What type of automatic control system is this (Fig. 8.2)?

(b) Sketch the shape of the graph of room temperature plotted against time, which would be obtained from this system.

(c) Starting from the basic block diagram, Fig. 8.1, deduce the particular block diagram of this system.

(d) What would the corresponding open-loop system consist of?

A CISTERN

Fig. 8.3 A section through a cistern

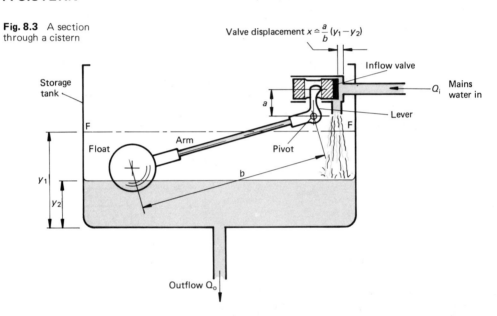

Valve displacement $x \triangleq \frac{a}{b}(y_1 - y_2)$

Fig. 8.3 shows a section through a cistern forming the header tank of a domestic hot water supply system. Its purpose is to control automatically the water level in the system to keep it from falling too far below the 'full' line FF. If the water level falls, due to water being drawn off from the system (outflow Q_o), the float, arm and lever rotate about the pivot, and allow mains water pressure to push the inflow valve away from its seating. This lets mains water in to replace the water being drawn off, and eventually to restore the level to FF, the level at which the float just closes the inflow valve.

SELF-TEST QUESTION 14 (Solution on p. 240)

(a) Starting from the basic block diagram, Fig. 8.1, deduce the particular block diagram of this system (Fig. 8.3).

(b) How could the reference input be altered?

(c) What would the corresponding open-loop system consist of?

A MECHANICAL SPEED CONTROL SYSTEM FOR AN ENGINE

The system shown in block diagram form in Fig. 7.3 on p. 157 is one possible way of automatically controlling engine speed. It was introduced at that point to illustrate the use of power amplification in an automatic control system. It is more usual, however, to use an all-mechanical system, incorporating a centrifugal governor. Fig. 8.4 illustrates the principle applied to a diesel engine.

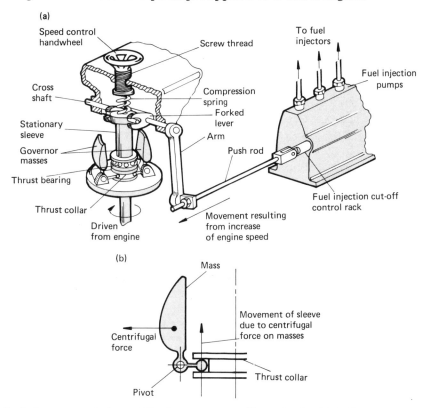

Fig. 8.4 (a) Typical arrangement of a centrifugal governor controlling a diesel fuel injection pump. (b) Action of masses

The feedback transducer is a pair of masses carried on a spindle driven from the engine. The masses are hinged to the spindle at their lower ends, and each has a 'toe' which projects at right angles into the groove in a thrust collar. As the engine speed increases, the masses pivot outwards under the action of centrifugal force, and their 'toes' lift the thrust collar as shown in Fig. 8.4(b). This movement is transmitted through a thrust bearing and a stationary sleeve to a forked lever carried by a cross-shaft. An arm on the cross-shaft, connected to the fuel injection pumps by a push-rod, moves a rack which alters the point at which fuel injection to each cylinder is cut off during the engine cycle. This alters the power output of the engine cylinders and hence controls the engine speed.

An equilibrium speed is reached at which the upward force exerted by the toes of the governor masses is balanced by the downward force of a compression spring acting on the top of the stationary sleeve. The equilibrium speed can be changed by altering the position of the upper end of the compression spring, by screwing it up or down by means of the speed control handwheel.

SELF-TEST QUESTION 15 (Solution on p. 241)

Draw the block diagram of this control system (Fig. 8.4).

CONTROL OF PLATE THICKNESS IN A STEEL ROLLING MILL

In a steel rolling mill, slabs of white hot steel pass between successive pairs of rollers (called *rolls*). Each pair rolls out the steel a little more, to make it longer and thinner so that at the end of the line, the material emerges as long strips of thin steel plate, ready for cooling and winding up into coils.

In a modern rolling mill the process is continuous, the slabs entering at a speed of about 2 metres per second, and the finished plate emerging at about 18 metres per second. Several automatic control systems are necessary for the success of this process: speed control, temperature control, plate width control, and so on. We are going to consider one such system, that controlling the thickness of the plate.

The thickness of plate at each stage in the rolling process is controlled by adjusting the gap between the rolls. To do this quickly enough, while exerting very large compressive forces on the plate, the rolls are screwed down by screw jacks driven by electric motors with ratings of several hundred kilowatts.

Power for these motors is available from the electricity supply grid, but there is a snag: the power available is a.c., and a.c. electric motors run in one direction only, at speeds tied to the supply frequency. They would be quite unsuitable for the precise control of roll position.

To overcome this, Ward–Leonard systems are usually used. In the Ward–Leonard system, a motor-generator set is used to generate d.c. from the a.c. supply, and this d.c. is used to power a work motor which does the actual job required.

The motor of the motor-generator set is an a.c. motor running at constant speed, and the output voltage of the generator it drives is controlled by varying the voltage applied to the generator field winding.

The work motor's field winding is supplied with d.c. from a constant voltage source, so the motor-generator output, applied to

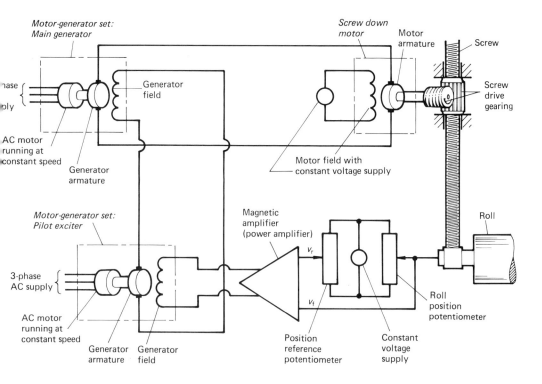

Fig. 8.5 Thickness-of-plate control in a rolling mill

the work motor's armature, controls the speed and direction of rotation of the output shaft.

The screwdown motor of a rolling mill has to transmit so much power that two stages of Ward–Leonard control are necessary, the d.c. supply to the field of the main motor-generator being provided by a smaller motor-generator, the *pilot exciter set*.

The field winding of the pilot exciter's generator is supplied from the output of a power amplifier, the two inputs of which are connected, one to a reference potentiometer and the other to a potentiometer whose wiper is set by the position of the upper roll. The output of the amplifier is proportional to the voltage difference between the two inputs, and its polarity (and hence, eventually, the direction in which the screwdown motor runs) depends on the plus or minus sign of that voltage difference.

Fig. 8.5 shows the relationship between all these elements of the roll position control system. The actual system includes some additional features, designed to improve its performance; these have been omitted for simplicity.

SELF-TEST QUESTION 16 (Solution on p. 242)
Draw the block diagram of this system (Fig. 8.5).

EXERCISES ON CHAPTER 8

1 Draw the basic block diagram of all automatic control systems which use closed-loop control with negative feedback.

2 Draw (a) the circuit diagram and (b) the block diagram of an on-off control system with which you are familiar, and give an account of its operation. Explain in detail how the error signal is generated, and the action of the on–off element of the system.

3 Fig. E8.1 shows the 'motor' and the position transducer for the pen carriage of a chart recorder. The 'motor' consists of a coil free to slide along a rod. The rod is part of a system of magnets, arranged to provide a permanent magnetic field at right angles to the rod throughout its length. Current through the coil sets up another magnetic field which propels the coil along the rod in the direction determined by the direction of current flow through the coil.

Fig. E8.1

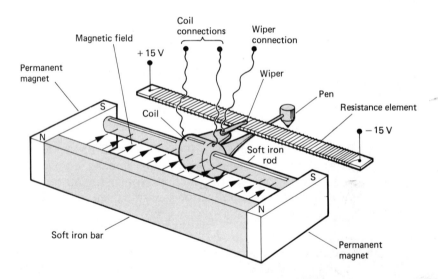

The position transducer is a straight-track potentiometer. The voltage picked off by its wiper is subtracted from the input signal voltage, and the + or − sign of the error operates an electronic switching unit to switch a current of constant magnitude through the coil in whichever direction gives motion tending to reduce the error to zero. Servos such as this, in which the maximum rate of correction is continually being applied in alternate directions, are called 'bang-bang' servos.

Draw the block diagram for this system as a closed-loop automatic control system.

4   Fig. E8.2 shows the principle of an automatic pressure regulator for use in a compressed air line. High-pressure air passes between the valve and its seating, then escapes upwards into the space underneath the diaphragm. From here it leaves the regulator through the low-pressure exit. The diaphragm hovers in equilibrium between the downward force of the spring on one side, and the regulated air pressure acting on the other side.

Fig. E8.2

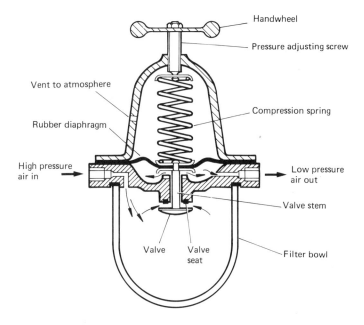

(a) Explain how the regulator keeps the downstream air pressure approximately constant in spite of large changes in demand for the pressure-regulated air.

(b) Draw the block diagram for this device as a closed-loop automatic control system.

5   The diagram, Fig. E8.3, represents a system for distributing liquid pesticide as evenly as possible from a crop-spraying vehicle. To give an even distribution, the flow rate must increase when the vehicle speeds up, and decrease when it slows down.

The transducer for vehicle speed is a tachogenerator driven from one of the vehicle's wheels. The flow rate is determined from the pressure of liquid in the pipeline, just before it leaves the nozzle. This is taken to a pressure transducer, which converts the pressure into a proportionate voltage.

The difference between the two voltages is input to an amplifier. The output of the amplifier drives a small electric motor. The motor rotates a control valve in the pipeline, through a 70:1 step-down gear train, to control the flow of the pesticide.

Fig. E8.3

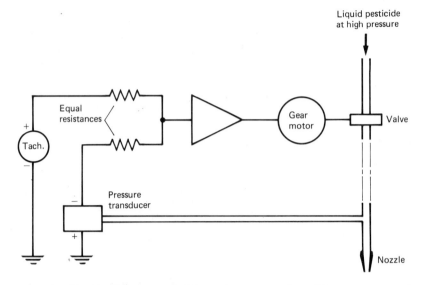

Draw the block diagram of this system, as a closed-loop automatic control system.

6   Fig. E8.4 shows the automatic focusing system of a modern slide projector.

Fig. E8.4

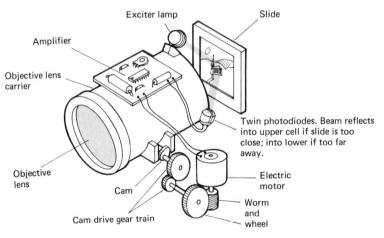

Focusing a projector is a matter of keeping the objective lens assembly at the correct distance in front of the slide to be projected. However, because of the intense heat radiated from the projection lamp, the slides buckle by varying amounts, so that without the automatic focusing system, the projector focusing would have to be readjusted each time another slide was inserted.

The automatic focusing system overcomes this by moving the objective lens assembly forwards or backwards until it is at the set distance from the centre of the slide transparency, no matter how far the slide has distorted. It does this by projecting a beam of light from an exciter lamp, carried by the objective lens assembly, at such an angle that it reflects off the surface of the transparency into a pair of photodiodes. Their electrical outputs go to the two inputs of an operational amplifier. The output of this, after power amplification, drives a small electric motor which, through step-down gearing, rotates a cam to move the objective lens forwards or backwards.

When a slide is correctly focused, the beam from the exciter lamp is reflected equally into both photocells, and their outputs cancel out. If the slide surface is behind or in front of the correct position, however, it deflects the reflected beam so that one photo-cell receives more light and the other less. The resulting difference in their outputs is translated by the system into movement of the objective lens assembly in the direction which will equalise the outputs of the photocells.

Draw the block diagram of this system as a closed-loop automatic control system.

7   Fig. E8.5 illustrates the working principle of an automatic tempera-ture control for the water supply to a shower, in which hot water and cold water are blended together to give the required shower temperature.

Fig. E8.5

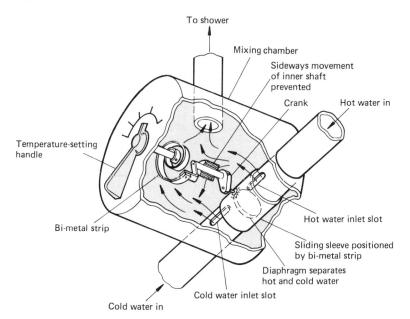

The shower temperature is set by a handle, which rotates one end of a spiral coil of bimetal strip. The other end of this strip rotates a crank, which moves a sleeve along the hot and cold water inlet pipe. The sleeve blanks off more or less of the slots through which hot water and cold water enter the mixing chamber, any movement increasing one inlet and decreasing the other.

Disturbances take the form of changes in the flow rate or temperature of one of the two water supplies. These give rise to changes in the temperature of the water passing the bimetal strip. The resulting change in its curvature rotates the crank and so moves the sleeve in the direction necessary to restore the output temperature to its original value.

Draw the block diagram of this system as a closed-loop automatic control system.

8   (a) Draw a schematic layout of a Ward–Leonard speed control system, and explain how it works.

(b) Give an example of an industrial process which uses a Ward–Leonard speed control system, and explain its particular advantages for that process.

Chapter 9

Some Standard Elements of Control Systems

COMPARISON ELEMENTS

THE DIFFERENTIAL LEVER

This is commonly used as an error detector in position control systems where the reference input and the controlled output are both linear displacements.

It consists of a lever suspended from three push-rods, as shown in Fig. 9.1. The two outer rods apply the input and feedback displacements, respectively, to the lever, while the remaining rod transmits the error signal as a displacement to the control elements of the system.

Fig. 9.1 Differential lever

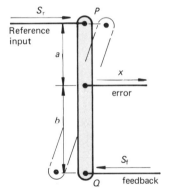

The input displacement and the resulting feedback displacement are arranged to act in opposite directions so that the error link connection returns to the same point in space when the control system has reduced the error to zero.

In Fig. 9.1, imagine that the reference input is given a displacement s_r so that the lever pivots about the point Q. Then, from the proportions of the lever:

$$\text{Error signal } x = \frac{b}{a+b} \times s_r$$

Now imagine that a feedback displacement s_f occurs in the opposite direction, pivoting the lever about P.

Then

$$x = -\frac{a}{a+b} \times s_f$$

The resultant error signal is the sum of these, that is:

$$x = \frac{b}{a+b} s_r - \frac{a}{a+b} s_f \qquad [1]$$

Of course, in a control system, the input and feedback displacements occur more or less simultaneously, not as two separate movements, but the effect is the same.

Thus the action of the differential lever is:

(a) it diminishes the input signal by a factor $b/(a+b)$;
(b) it diminishes the feedback signal by a factor $a/(a+b)$;
(c) it acts as an error detector with negative feedback.

It therefore appears in three places in a control system block diagram, as shown in Fig. 9.2.

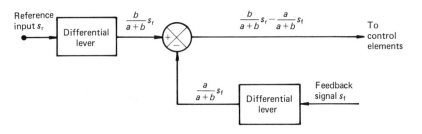

Fig. 9.2 Representation of a differential level in a control system block diagram

When a differential lever is used in a positional servo system, such as the hydraulic servo copying system for a lathe shown in Fig. 9.9 on p. 182, the feedback link continues to move until the spool valve returns to the closed position — that is, until $x = 0$. Then from equation [1],

$$s_f = \frac{b}{a} s_r \qquad [2]$$

If a and b are equal, we have a *symmetrical differential lever* and equation [2] becomes $s_f = s_r$. That is, for practical purposes, the input and feedback displacements are equal, but take place in opposite directions.

THE DIFFERENTIAL POTENTIOMETER AND THE SUMMER AMPLIFIER

The potentiometer has already been described on pp. 78-9, and used in the control systems shown in Figs. 7.3 and 8.5. In position control systems where the error signal is required in the form of a voltage, it is very convenient to be able to use one potentiometer to provide the reference voltage and another to provide the feedback voltage. Error detection with negative feedback occurs automatically if one of the potentiometers has its connections to the d.c. supply reversed, as shown in Fig. 9.3. If the wipers of the two potentiometers keep in step, they produce voltages of opposite polarity, which are added to provide the error signal. This is done by feeding the respective voltages through equal resistances R to the input of an operational amplifier as shown in Fig. 9.3. The operational amplifier is then being used for *summing*, as shown on p. 61 (Fig. 3.14).

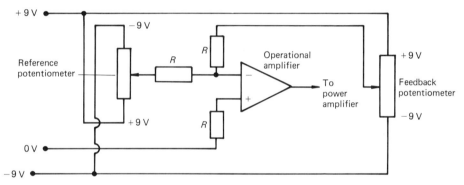

Fig. 9.3 Using reference and feedback potentiometers with opposite supply connections to obtain negative feedback by addition. The ± 9 V is a typical power supply voltage

We have already met a similar application of two potentiometers in the XY plotter servo circuit, shown in Fig. 4.6 on p. 98.

THE SYNCHRO

A synchro consists of:

(a) *a rotor*, or rotating part, consisting of an armature wound with a single coil, and

(b) *a stator*, or stationary housing, containing three identical coils positioned 120° apart around the housing, as shown in Fig. 9.4.

Synchros are used in pairs. A pair of synchros can be used:

(a) to form a passive, remote-reading angular position indicator, or

(b) as input position transducer and feedback transducer in an angular position control servomechanism.

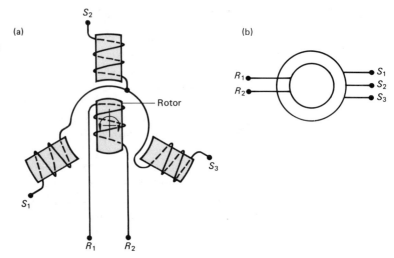

Fig. 9.4 Synchro: (a) wiring diagram; (b) symbol used in schematic diagrams

The first of these two applications does not concern us, as the synchros do not *interact* with an automatic control circuit. The second application is the one in which we are interested, because in this arrangement the synchros act as a comparison device. The electrical circuit is shown in Fig. 9.5.

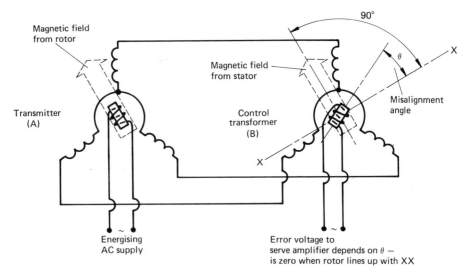

Fig. 9.5 Pair of synchros acting as an error detector

An a.c. voltage is applied to the rotor winding of the transmitter synchro (A), setting up a pulsating magnetic field. This induces voltages in the stator coils of A. These voltages are applied to the corresponding coils of the control transformer (B) and set up a magnetic field parallel to A's. This sets up an a.c. voltage in the rotor winding of B, with an amplitude which depends on the angular position of that rotor. The amplitude becomes zero, when B's rotor is exactly at right angles to A's.

Thus the two synchros act as the error detector in a position control servo. The output of B's rotor winding is an a.c. error signal, with magnitude and in-phase or anti-phase state depending on the misalignment angle. Fig. 9.6 shows the block diagram of a position control servo using a pair of synchros as the error detector, and Fig. 9.7 shows the actual arrangement of the components. The a.c. error signal has to be converted to a d.c. signal by means of a phase-sensitive detector. (We have already met this

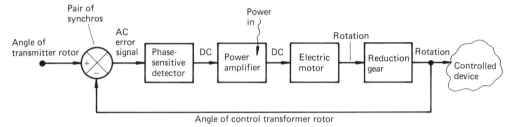

Fig. 9.6 Block diagram of a position control servo using synchros as the error detector

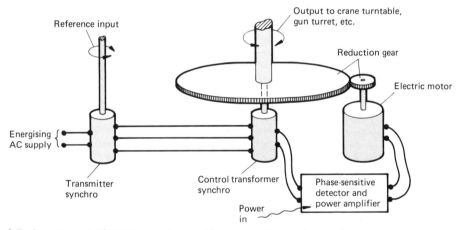

Fig. 9.7 Arrangement of components in a position control servo using synchros

device when we studied the LVDT — Chapter 2, pp. 41-2.) The d.c. signal, after power amplification to provide the necessary current, drives an electric motor which powers the controlled device.

CONTROL ELEMENTS

THE SPOOL VALVE

The *spool valve* (also called a *servo valve*, a *pilot valve* or a *relay valve*) is a cylindrical steel rod, machined to leave *lands* of full diameter with recessed sections in between. These lands are ground and lapped to fit the bore of the valve body. By blocking or uncovering holes (*ports*) in the valve body, they control the admission of hydraulic fluid from a high-pressure supply to one side of an actuator piston, and allow fluid from the other side of the actuator to escape to the fluid reservoir.

Fig. 9.8 Spool valve

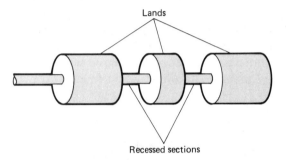

Because equal and opposite pressure forces act on the faces at either end of a recessed section, they cancel out, and thus there is no net end load on the valve.

Fig. 9.8 shows a typical spool valve, and Fig. 9.9 shows an application of a spool valve in which it is used as the control element in a copying device for a lathe.

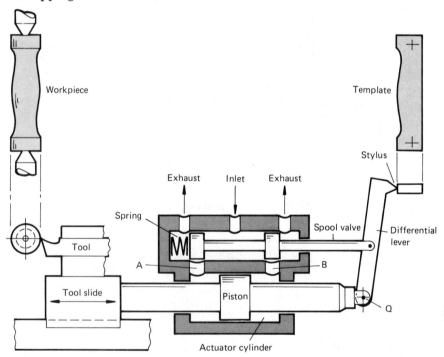

Fig. 9.9 Hydraulic servo copying system for a lathe

The whole device is carried by the saddle of the lathe. As it is traversed along the lathe bed, the stylus, which forms part of a symmetrical differential lever, traverses along a fixed template of the profile to be machined.

Suppose the stylus moves to the right, in following the profile. The differential lever hinges about pivot pin, Q, moving the spool valve to the right. This admits high pressure hydraulic fluid into the actuator cylinder via port, B, while fluid on the other side of the piston is able to escape via port, A, and the left-hand exhaust port to the reservoir. The piston therefore moves to the left. In doing so, it returns the spool valve to the central position, blanking off both port A and port B, and thus locking the piston between two (almost) incompressible volumes of hydraulic fluid.

Fig. 9.10 Block diagram of the system shown in Fig. 9.9

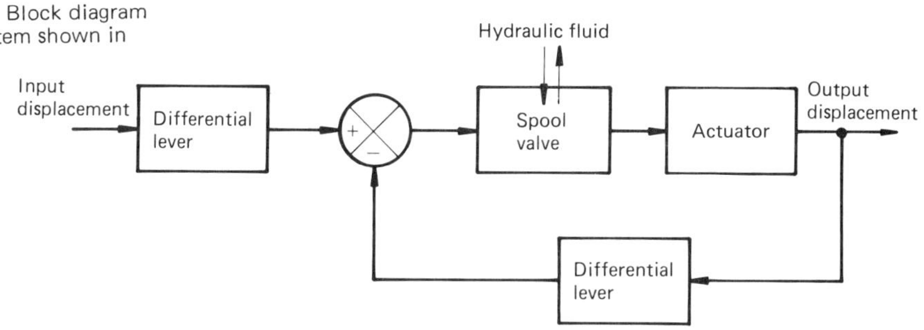

Because the spool valve connection to the differential lever is equidistant between the stylus and pivot pin, Q, the movement of the piston is equal and opposite to the movement of the stylus, as explained on p. 178, and the profile of the template is exactly duplicated in the profile of the workpiece.

THE FLAPPER AND NOZZLE

If a fluid is discharged through a nozzle against a perpendicular flat surface, as shown in Fig. 9.11(a), a back pressure is set up, which depends on the gap between the nozzle and the surface, as shown in Fig. 9.11(b). This is the principle on which a measuring device, the pneumatic comparator, is based. To prevent the back pressure from being masked by the supply pressure, the flow from the supply is limited by a restriction (usually, a plug with a small hole drilled through it).

Fig. 9.12 shows the same principle used in a control transducer, converting an input displacement into a change of back pressure. The device is called a *flapper-nozzle valve*. The flapper may be a springy steel strip, rigidly fixed at the root, or it may be a rigid member, pivoted as shown. The fluid may be compressed air or hydraulic fluid.

The flapper is usually used with two nozzles, so that as it moves closer to one nozzle it moves further away from the other. The output of this arrangement is a pressure difference which can be used to operate a piston. Fig. 9.13 shows the flapper-nozzle used

Fig. 9.11 (a) Principle of the pneumatic comparator. (b) relationship between back pressure and gap

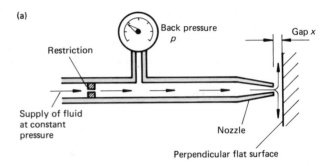

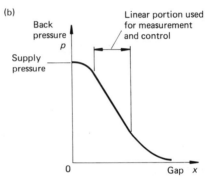

as an intermediate stage in a servo valve which has an electrical input and two stages of hydraulic amplification.

The input signal is a current supplied by a power amplifier, the direction of the current depending on the plus or minus sign of the error signal in the closed-loop control system of which the servo valve forms part.

The current flows through a coil surrounding a soft-iron 'tail' on the flapper, magnetising it so that one end, between the poles of a permanent magnet, is pulled towards one or other of the poles of the magnet by a force proportional to the input current.

Fig. 9.12 Flapper-and-nozzle control transducer

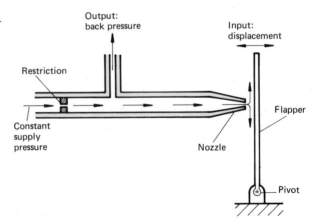

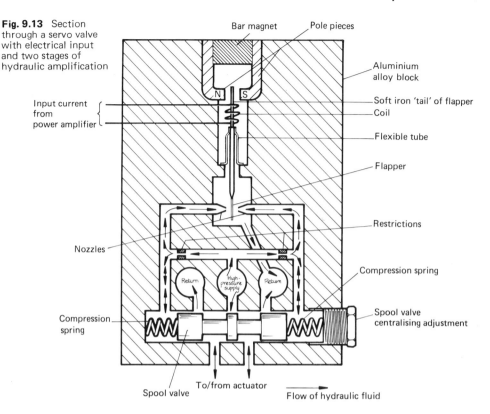

Fig. 9.13 Section through a servo valve with electrical input and two stages of hydraulic amplification

The flapper is, in effect, pivoted near its mid-point, by being mounted in a flexible tube. This allows it to pivot slightly about the root of the tube, while at the same time the tube prevents the hydraulic fluid emitted by the nozzles from escaping.

The flapper is situated between the two nozzles and takes up a position where it is in equilibrium between the moment of the magnetic forces at one end, and an opposing moment applied by jets from the nozzles at the other end.

The spool valve can slide freely in its bore in the servo valve body. It is centralised by compression springs pushing on each end. The back pressures from the nozzles are also applied to the respective ends of the spool valve, and so the spool valve, in its turn, takes up a position where it is in equilibrium between the difference in forces due to back pressure from the nozzles, and the difference in forces from the compression springs.

Thus for a given input current there is:

(a) a proportional displacement of the flapper, and hence

(b) a proportional displacement of the spool valve, and hence

(c) a proportional flow of hydraulic fluid to operate an actuator.

The original spool valve, shown in Fig. 9.9, converts linear displacement into oil flow rate. The servo valve of Fig. 9.13 converts current into oil flow rate. As we have seen, it is more complicated, but the extra complication is justified by the fact that it is controllable by an electrical error signal instead of a mechanical one, and thus the response of the system can be 'shaped' by taking the input signal through an electronic controller.

THE PROCESS CONTROL VALVE

Process control valves are used in systems where the flow of a gas or liquid is to be continuously regulated by automatic control. They are usually operated by compressed air, so that the valve opening varies with the air pressure. Fig. 9.14 shows a section through a typical process control valve.

Fig. 9.14 (a) Section through a process control valve. (b) Symbol used in schematic diagrams

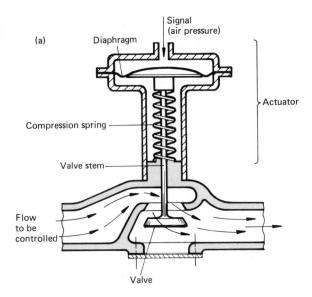

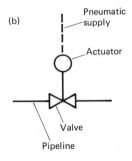

The controlling air pressure is varied between zero and about 130 kN/m² (gauge). It acts on a rubber diaphragm, exerting a large force which compresses the return spring and opens or closes the valve. The valve opening is determined by the spring deflection necessary to make the force from the return spring equal to the force on the diaphragm.

The valve shown in Fig. 9.14 is an 'air-to-open' valve — that is, the controlling air pressure pushes the valve *off* its seating.

Fig. 9.15 shows an 'air-to-close' valve, in which the controlling air pressure pushes the valve *towards* its seating. In deciding which type to use for a particular application, the designer usually works on the principle of 'fail–safe', and chooses the type which would leave the plant in a safe condition if the air supply failed. Thus if, for example, the valve controls the fuel supply to a furnace, an air-to-open valve would be used, so that in the event of loss of air pressure, the fuel would be cut off to prevent overheating. If, however, the valve regulates the flow of cooling water to, say, a heat exchanger, an air-to-close valve would be used, so that in the event of air failure the valve would open fully, again preventing damage from overheating.

Fig. 9.15 Air-to-close valve

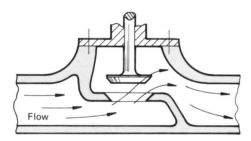

Flow

Although air pressure on the diaphragm is used to set the distance the valve moves off its seating, the actual distance it moves is influenced also by the rate of flow of fluid past the valve, and its direction. In Figs. 9.14 and 9.15, the direction of flow shown tends to push the valve further open, exerting a force which is quite considerable — as much as 5 kN on a valve of 100 mm diameter, if the pressure drop across the valve is 600 kN/m². If the flow is in the opposite direction, it tends to close the valve, with a similar force.

A further complication is the inertia effect of the fluid whose flow is being controlled. A steady flow of a dense liquid, such as water, through a long pipe, has considerable momentum, and when we try to alter the flow rate, the compressibility of the fluid and the elasticity of the piping can cause oscillations of flow which may build up violently. An example of this is the steady hammering or buzzing (*water hammer*) which is sometimes emitted by domestic water supply systems when a tap is turned on or off. This is caused by an automatic control system (the cistern, described in Chapter 8) going unstable due to the mass of the water and the elasticity of the supply system to which it is connected.

If the flow is in the direction tending to push the valve open, as shown in Figs. 9.14 and 9.15, flow-induced oscillations of the valve tend to die out quickly, but if the flow is in the opposite direction, tending to close the valve, it is much more likely to oscillate with increasing amplitude so that the system becomes unstable.

One way of (almost) eliminating the effects of the drag of the fluid on the valve is to use a double-seated valve, as shown in Fig. 9.16, so that one of the valves tends to be opened by the fluid while the other tends to be closed. Thus the drag forces on the valve plugs almost entirely cancel out and the valve is much more stable.

The disadvantage of this type of valve is that it is difficult to ensure that both plugs seat simultaneously, so it is not suitable where the valve must be capable of completely sealing off the flow.

Fig. 9.16 Section through a double-seated valve

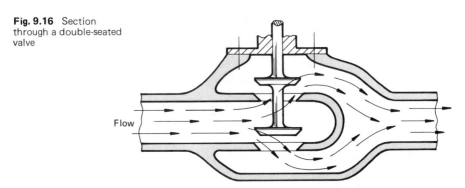

Flow

LINEAR HYDRAULIC ACTUATORS (HYDRAULIC CYLINDERS)

The principle of the linear hydraulic actuator has already been illustrated in Fig. 9.9, where an actuator with a double-ended piston rod is shown, controlled by a spool-valve and differential lever. Usually, however, the piston rod extends from one end of the cylinder only. The actuator may be either *single-acting* or *double-acting*, and where the momentum of the piston and its load may become so great as to damage the actuator by impact at the end of the stroke, *cushioning* can be provided. These terms are explained below.

A single-acting cylinder has only one connection for hydraulic fluid, and can apply force in only one direction, as shown in Fig. 9.17. The piston is returned to its starting point by the load, when the fluid pressure is released. Hydraulic jacks for motor vehicles work on this principle. Where the load is not able to return the piston, an internal spring provides the return force, as shown in Fig. 9.18.

Fig. 9.17 Single-acting cylinder

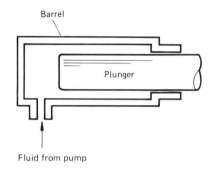

Barrel

Plunger

Fluid from pump

Fig. 9.18 Single-acting cylinder with return spring

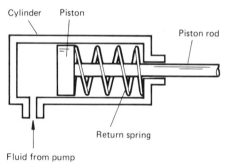

Cylinder   Piston

Piston rod

Return spring

Fluid from pump

A double-acting cylinder has connections for hydraulic fluid at both ends of the cylinder as shown in Fig. 9.19. These act as inlet and exhaust connections alternately, so a double-acting cylinder can apply force in either direction. The force it can apply on the return stroke is less than that on the outward stroke, because on the piston rod side, the area of the piston is reduced by the cross-sectional area of the piston rod. For the same reason, the speed on the return stroke is greater, for a given fluid delivery rate, because the volume to be filled with fluid on the piston rod side is less.

Fig. 9.19 Double-acting cylinder. The inlet and return ports exchange their roles for the return stroke

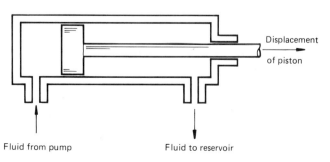

Displacement

of piston

Fluid from pump   Fluid to reservoir

Cushioning of the piston can be arranged at either end of its travel. The principle is shown in Fig. 9.20. As the piston approaches the end of its travel, the tapered end of the plunger enters the hole in the cylinder head. This restricts the area through which fluid can pass out to the reservoir. The parallel part of the plunger follows, closing off the normal outlet for fluid completely. The only outlet then is past the adjustable restriction, as shown by the arrows. The resulting rise in pressure on the left-hand side of the piston decelerates it to a cushioned stop.

The spring-loaded ball valve shown in the upper part of the diagram (Fig. 9.20) is necessary to allow unrestricted passage of high pressure fluid at the beginning of the next outward stroke of the piston, until the plunger has left its bore in the cylinder head.

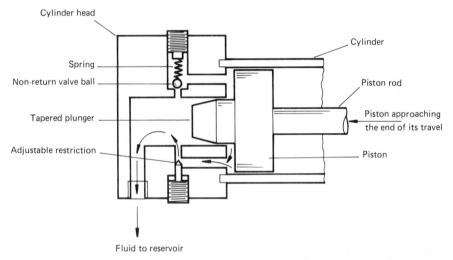

Fig. 9.20   Section through a cylinder and cylinder head to show the principle of end position cushioning

ROTARY HYDRAULIC ACTUATORS

These are typically used for controlling plug valves or butterfly valves in pipelines, where 90° of rotation is sufficient to fully open or close the valve. Other applications include indexing or inverting a work-piece, and tilting or dumping the contents of skips. They are intended for rotations of less than 360°; for greater rotations than this, one would use a hydraulic motor.

The simplest type of rotary actuator is shown in Fig. 9.21. The rotating vane rotates the shaft when hydraulic fluid is pumped into the space on one side of it, the fluid on the other side being returned to the reservoir through the other hydraulic connection. About 280° of rotation is available from this type of actuator. A similar type has two stationary barriers, diametrically opposite each other, and two rotating vanes also diametrically opposite

Fig. 9.21 Vane-type
rotary actuator

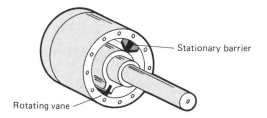

Stationary barrier

Rotating vane

each other. This reduces the available angle of rotation to about
100°, but gives twice the torque for a given size and pressure. Also,
the forces on the shaft are balanced, giving a pure torque instead
of a moment.

Another type of rotary actuator is shown in Fig. 9.22. It is, in
effect, a double-acting hydraulic cylinder which has an internal
plunger in place of the piston. Rack teeth on the plunger rotate
the actuator shaft by means of a pinion integral with the shaft.
This type of actuator can provide rotations up to about 360°.

Fig. 9.22 Rack-and-
pinion type rotary
actuator

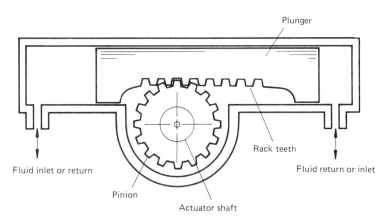

Plunger

Rack teeth

Fluid inlet or return

Fluid return or inlet

Pinion

Actuator shaft

D.C. AND A.C. SERVO-MOTORS

The main features of the ideal electric motor to power a servo
control system would be:

(a) high starting torque;

(b) wide speed range;

(c) ability to run equally well in either direction;

(d) low moment of inertia of the armature for quick acceleration and deceleration;

(e) self-braking ability, to bring the armature to rest quickly when
the controlling current is switched off, to reduce overshoot.

Items (a) and (d) work against each other and it is necessary to
compromise between them.

D.C. SERVO-MOTORS

A d.c. electric motor consists electrically of two components:

(a) A rotating armature carrying coils connected to the segments of a commutator.

(b) A casing which provides a stationary magnetic field within which the armature pulls itself round.

In very small electric motors the casing is itself a permanent magnet with internal magnetic field, so that the only means of control is the current through the armature.

Large d.c. servo-motors operate on the same principle, but the stationary field is provided by electro-magnets. Their coils (the *field windings*) are usually supplied from a constant voltage source, the speed and direction of rotation of the armature being controlled by the current through the armature coils only.

Most d.c. servo-motors, however, operate on the split-field principle, illustrated in Fig. 9.23. The armature is fed from a constant current source, and its speed and direction of rotation are controlled by the resultant magnetic flux set up by the two halves of the field winding. This is fed from a 'push-pull' type of amplifier: in effect, two identical power amplifiers 'back-to-back', with outputs at C and D and a common return connection at E. When equal currents flow as shown in Fig. 9.23, the magnetic effects of the two halves of the field winding cancel out, and the armature is stationary. An error signal unbalances this arrangement, causing a greater current to flow through one half of the field winding and a smaller current through the other half. The motor then runs in the direction determined by the greater current flow. The torque-speed-current relationship is shown in Fig. 9.24.

Fig. 9.23 The essential features of a split-field d.c. servo-motor

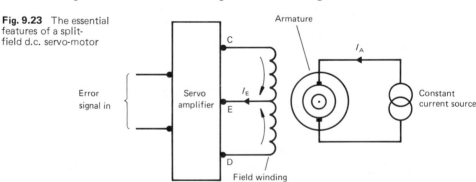

The field windings usually have many turns because the available current is small. Their inductance is therefore considerable; thus sudden changes in field current produce high voltage surges. A well designed split-field servo-motor should therefore have surge protection shunts in parallel with the field windings.

Fig. 9.24 Torque/speed relationship of a split-field d.c. servo-motor

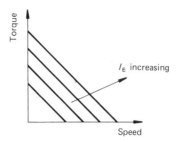

A d.c. servo-motor usually has a long thin armature, to give the best torque compatible with low moment of inertia.

A.C. SERVO-MOTORS

Most a.c. servo-motors are *two-phase induction motors*. In this type of motor there are two sets of field windings, producing magnetic fields at 90° to each other.

In an ordinary small two-phase industrial motor, the two windings are supplied from the same a.c. source but with the current in one winding 90° out-of-phase with the current in the other. This produces, in effect, a rotating magnetic field, which induces eddy currents in a system of conductors (the 'squirrel cage') embedded in the armature (now called the *rotor*). The magnetic field from these eddy currents is attracted to the rotating magnetic field, and so pulls the rotor after it. Thus there is no electrical connection to the rotor, in the form of commutator or slip rings, at all.

An industrial two-phase induction motor runs at an approximately constant speed, in one direction only. Fig. 9.25 shows how the principle of the two-phase motor is adapted to make it suitable for use as a servo-motor. One of the two field windings, the reference winding, is supplied with a constant a.c. voltage, V_R. The other, the control winding, is supplied from an a.c. amplifier which gives a current I_C at the same frequency but either leading or lagging behind the reference voltage by 90° — this *lead* or *lag* sets the direction of rotation of the rotor. The speed of rotation is determined by the amplitude of this control current.

Fig. 9.25 The essential features of a two-phase induction type a.c. servo-motor

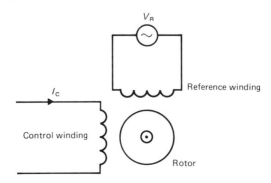

The rotor core is designed to have low magnetic reluctance and high electrical resistance, to give the high starting torque necessary for fast response in a servo-motor.

The torque-speed-current relationship is shown in Fig. 9.26. Notice that the curve for zero control current ($I_C = 0$) passes through the origin and has negative slope. This indicates that if the control current is shut off while the rotor is still turning, a definite braking torque is produced by the eddy currents generated in the rotor conductors — a useful property in a servo-motor.

Fig. 9.26 Torque-speed relationship of a two-phase induction type a.c. servo-motor

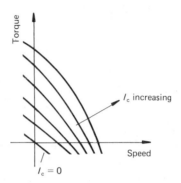

EXERCISES ON CHAPTER 9

1 (a) A lever is commonly used to magnify translational position or to perform the computation of subtraction. Explain how the lever is used in these two cases.

(b) Describe and compare a jewel bearing, a journal bearing and a fulcrum as a means of pivoting the lever.   (I.Q.A.)

2 (a) Sketch a differential lever.

(b) Draw the arrangement of blocks representing a differential lever in the block diagram of an automatic control system.

(c) A differential lever has the following dimensions:
120 mm from the input link pivot to the error link pivot;
80 mm from the error link pivot to the feedback link pivot.

If the input displacement is s_r and the feedback displacement is s_f, calculate the error in terms of s_r and s_f.

3 In a position control servo, the input is a displacement applied to the wiper of a potentiometer, giving an input signal in the form of a voltage. The corresponding output displacement moves the wiper of a similar potentiometer, giving a feedback signal in the form of another voltage. Show how the two potentiometers would be connected to the d.c. power supply, and to each other, so that an error signal corresponding to the difference of the two voltages is obtained.

4   Draw the electrical circuit diagram of a synchro, showing the stator and rotor windings and the electrical connections.

5   (a) Draw the electrical circuit diagram of a pair of synchros connected together as an error detector. The diagram should show the stator and rotor windings, the necessary connections between the two synchros and the electrical input to and output from the system. Indicate, by means of an arrow, the direction of the magnetic field in each synchro. Label the *transmitter* and the *control transformer*.

   (b) Referring to your diagram, explain the operating principle by which the system produces an error signal related to the misalignment between the rotors.

6   (a) Sketch the arrangement of components in a position control servo which uses a pair of synchros as the error detector. Show the electrical connections between the components, and any electrical inputs necessary.

   (b) Draw the block diagram of the system.

7   Compare the advantages and disadvantages of:

   (a) a pair of potentiometers,

   (b) a pair of synchros,

   for producing the error signal in a position control servo.

8   (a) Sketch the arrangement of a hydraulic actuator consisting of a hydraulic cylinder with a spool valve to control the position of the piston, the spool valve and piston being linked by a differential lever to form an automatic position control system.

   (b) Draw the block diagram of this automatic control system.

9   (a) Make a sketch to show the basic arrangement of a 'back pressure' type of pneumatic comparator.

   (b) By means of a sketched graph show the relationship between (i) the back pressure and (ii) the gap through which air escapes. Indicate the part of the graph which is used for measurement or control purposes.

   (c) Sketch the flapper-and-nozzle arrangement corresponding to (a) which is used in control systems.

10   Show, by sketch and description, how a flapper valve may be used to control the position of a spool valve so that large flow rates of compressed air or hydraulic fluid may be controlled by very small displacements of the flapper.

11   (a) Sketch the general arrangement of a pneumatically operated process control valve, and explain how it works.

   (b) Show the usual symbol for such a valve on a plant layout diagram.

12   (a) Explain, with the aid of sketches, the difference between an 'air-to-open' and an 'air-to-close' process control valve.

     (b) Which of the two types would you choose to control:

       (i)   a flow of inflammable gas into a pressure vessel used to store the gas at constant pressure;

       (ii)   a supply of air for a continuous combustion process in which it is desirable to maintain an ideal fuel/air ratio under varying rates of fuel usage, but essential to avoid soot formation from incomplete combustion?

     In each case explain the reason for your choice.

13   With the aid of a sketch, explain how the stability of an automatic flow-rate control system may be affected by the direction of the flow past the plug of the control valve.

14   Sketch a section through a double-seated process control valve, and discuss the advantages and disadvantages of this type of valve compared with the single-seated type.

15   Explain with the aid of a sketch how the piston of a linear hydraulic actuator may be cushioned at the end of its travel inside the hydraulic cylinder.

16   Sketch and describe the principle of operation of two types of rotary hydraulic actuator.

17   List the desirable features of an electric motor to power a servo control system.

18   (a) Explain with the aid of a diagram the operation of a split-field d.c. servo-motor controlled by a 'push–pull' amplifier.

     (b) Why might the field windings need surge protection shunts?

19   (a) Show diagrammatically the essential features of a two-phase induction type a.c. servo-motor.

     (b) Explain how it may provide a positive braking torque.

Chapter 10

System Response

SPRING-MASS SYSTEM WITH VIBRATION DAMPING

Fig. 10.1 shows a mass able to slide freely along a smooth surface. The motion of the mass is modified by the attached dashpot piston which applies a force proportional to its velocity, in the opposite direction to the motion. The mass is attached to a spring, and input displacements are applied to the other end of the spring. The input and output displacements, x_i and x_o, are measured as shown from a datum positioned so that $x_i = x_o = 0$ when the system is at rest.

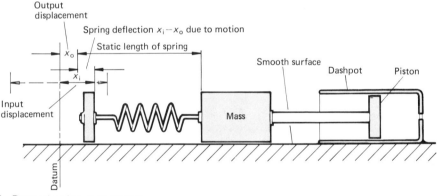

Fig. 10.1 Damped spring-mass system

Because the force exerted by the spring is proportional to the spring deflection, and this is obtained by subtracting its static length from its length at any instant, the static length of the spring acts as an inherent feedback device, feeding back the displacement of the mass for subtraction from the input displacement.

The spring–mass system can therefore be considered as an automatic control system, with block diagram as shown in Fig. 10.2.

Fig. 10.2 Spring-mass system considered as an automatic control system

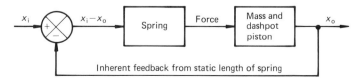

The fact that a damped spring–mass system is a member of the family of automatic control systems enables us to apply spring–mass vibration theory to the understanding of control system behaviour. The treatment which follows is derived by considering the spring–mass system, but applies equally to automatic control systems.

RESPONSE TO A STEP INPUT

Imagine that we can adjust the size of the hole through which air can enter or leave the dashpot in Fig. 10.1. Let us apply a step input to the system, that is, a sudden increase of dimension x_i.

If the hole in the dashpot is small, so that it applies a large force to the mass when it moves, the mass will move sluggishly to the equilibrium position where x_o again equals x_i, as shown in Fig. 10.3(a). The system is said to be *overdamped*.

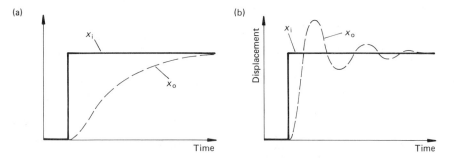

Fig. 10.3   System response: (a) over-damped; (b) under-damped

If the hole is large, so that the dashpot applies only small forces, the mass will move so quickly that it overshoots the equilibrium position, maybe several times, before coming to rest with x_o equal to x_i. That is, there will be a decaying oscillation as shown in Fig. 10.3(b), and the system is said to be *underdamped*.

By trial and error, we could adjust the damping so that the mass returned to the equilibrium position in the shortest possible time *without* overshooting. The system would then be *critically damped* — i.e. on the borderline between overdamped and underdamped.

The behaviour of a spring–mass system depends on the *damping ratio* of the system. The damping ratio, denoted by the Greek letter *zeta*, is calculated as

$$\zeta = \frac{\text{Actual damping force per unit velocity}}{\text{Damping force per unit velocity which gives critical damping}}$$

For critical damping, $\zeta = 1$. If the system is overdamped, ζ is greater than 1, and if it is underdamped, ζ is less than 1. Fig. 10.4 compares the step input response of the system for various values of ζ.

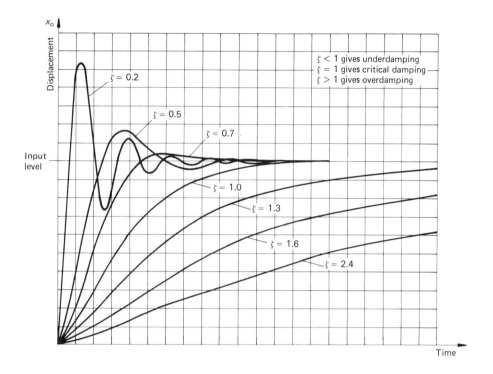

Fig. 10.4   System response to a step input for various values of damping ratio ζ

For a spring–mass system damped by a dashpot, the damping ratio, ζ, can be calculated from the mass, the spring rate, and the damping constant (i.e. the damping force per unit velocity) of the dashpot, all of which can be measured. The response of an automatic control system also follows the curves of Fig. 10.4, but in this case, ζ is more difficult to determine analytically. However, if the system has been built, and is found to oscillate in response to a step input, ζ can be calculated from the ratio of the amplitudes of successive half-cycles of the oscillation, as shown on p. 205.

Although critical damping ($\zeta = 1$) gives the fastest response to a step input *without overshoot*, for practical purposes an even faster response is given by a damping ratio of $\zeta = 0.7$, because, as can be seen from Fig. 10.4, between $\zeta = 1$ and $\zeta = 0.7$, the overshoot is small enough to be negligible. Thus instruments such as moving-coil meters are usually designed to have a damping ratio of about 0.7 so that they indicate their true reading as quickly as possible.

SOME DEFINITIONS

At this point it will be helpful to define the terms by which the various kinds of step input response shown in Fig. 11.4 may be specified. Some of these definitions have appeared already, in earlier chapters. Fig. 10.5 collects them all together.

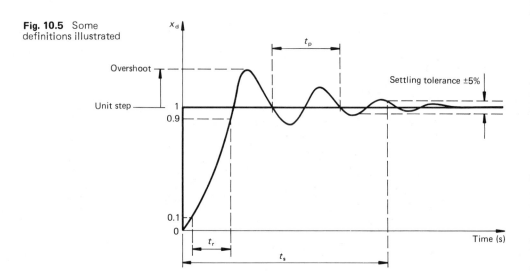

Fig. 10.5 Some definitions illustrated

Overshoot: this applies to underdamped systems only. It is the maximum amount by which the actual response exceeds the final steady output value.

Rise time: the time, t_r, needed for the response to rise from 10% to 90% of its final value.

Settling time: the time, t_s, needed for the response to reach its final value, within some specified tolerance. Fig. 10.5, as an example, shows it as the time taken for the output to settle within ± 5% of the final value.

Periodic time: the time, t_p, for one complete cycle of oscillation. The easiest way to measure it is to measure the time interval between *alternate* crossings of the final output level.

Frequency: the number of cycles of oscillation per second. This is calculated as:

$$\text{Frequency (Hz)} = \frac{1}{\text{Periodic time}}$$

Time constant: We have already met this in connection with piezo-electric transducers, on p. 39. It gives us a measure of how quickly a first-order system will reach (for practical purposes) its final value. The main points about it are re-stated in Fig. 10.6.

They are:

The time constant, τ (tau), is the time it takes for the measured quantity to reach 63.2% of its final value (or, looking at it in another way, be within $100 - 63.2 = 36.8\%$ of it).

After 3 times the time constant, the measured quantity will have reached 95% of its final value — that is, it will be within 5% of it.

After 5 times the time constant, the measured quantity will have reached 99% of its final value — that is, it will be within 1% of it.

The time constant is also the time it would take to reach its final value if the rate of change of the measured quantity could be kept constant, at the value it has at any particular instant. Thus τ can also be found from a tangent to the curve at any point, as shown in Fig. 10.6.

Fig. 10.6 The main features of time constant

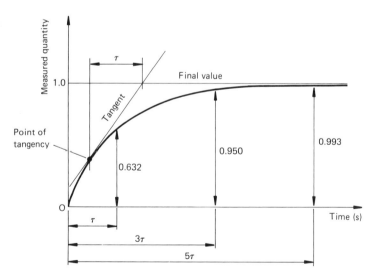

This concept can be applied to the system response curves of Fig. 10.4. It can be seen that they all have reverse curvature close to the origin, so none is a pure first-order system curve. However, with the overdamped curves we can take an artificial origin at, say, 30% of the unit step level, and thus obtain a value for the time constant from that point.

For the response curves of the underdamped systems, the time constant gives a measure of the rate of decay of the oscillation. Fig. 10.7 shows the typical response of a lightly damped system to a step input. The 'envelope' which forms the boundary of the oscillation is a pair of pure exponential curves, to which the time constant treatment can be applied.

Fig. 10.7 The envelope of a lightly damped oscillation

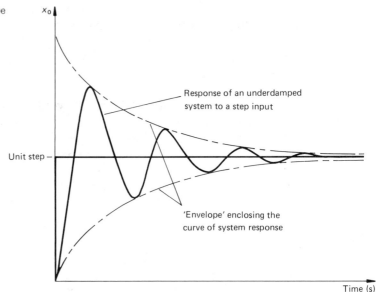

Response of an underdamped system to a step input

Unit step

'Envelope' enclosing the curve of system response

Time (s)

The time constant of these exponential curves can be calculated from

$$\tau = \frac{1}{\zeta \omega_n}$$

where ω_n is 2π times the undamped natural frequency. (See pp. 203–4.) Actually, if the damping is light, it will not matter very much if we use the actual frequency — that is, the damped natural frequency.

Once we have calculated the time constant of the envelope curves, we can say that within 3 times the time constant, the response will have settled to within $\pm 5\%$ of the final value, and within 5 times the time constant it will have settled to within $\pm 1\%$.

UNDAMPED NATURAL FREQUENCY

If we have a mass suspended from a spring (or supported by a springy structure) and we displace the mass from its equilibrium position so that the spring deflects, we do work which becomes stored as strain energy in the spring. If we now release the mass, the strain energy becomes converted to kinetic energy as the mass accelerates back to the equilibrium position. Because of its inertia the mass overshoots the equilibrium position until the spring is deflected an equal amount in the opposite direction and the kinetic energy reconverted to strain energy. The process continues indefinitely, the mass oscillating (vibrating) between the two extreme positions, and the energy, put into the system by the original displacement, continually being converted from strain energy to kinetic energy and vice versa. Fig. 10.8 shows the motion of the system.

Fig. 10.8 Vibration of a spring–mass system

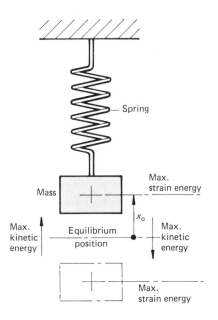

The system shown in Fig. 10.8 is equivalent to that in Fig. 10.1 except that there is no damping and no input. We can therefore represent it by the block diagram, Fig. 10.9, which is Fig. 10.2 suitably modified.

Fig. 10.9 Block diagram of an undamped spring–mass system with zero input

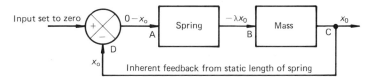

Let the mass be displaced a distance x_0 from its equilibrium position and released.

Then at C in Fig. 10.9, the output of the system is x_0.

Via the inherent feedback link, this is subtracted from the input displacement ($=$ zero), to give an error signal of $-x_0$ at A.

The input to the spring is displacement, and its output is force. The two quantities are related by the spring rate:

$$\lambda = \frac{\text{Force}}{\text{Displacement}}$$

So an input displacement of $-x_0$ applied to the spring exerts an output force at B of $-\lambda x_0$ on the mass. (The negative sign indicates that the force acts in the opposite direction to original direction of x_0 at C.)

The input force of $-\lambda x_0$ on the mass causes an acceleration determined by the relationship:

$$\text{Force} = \text{Mass} \times \text{Acceleration}$$

$$\therefore \qquad \text{Acceleration} = \frac{\text{Force}}{\text{Mass}}$$

$$= -\frac{\lambda}{m} \times x_0$$

This complies with the mathematical conditions for simple harmonic motion, the acceleration being of opposite sign and proportional to the displacement.

Hence the mass will vibrate with a natural frequency equal to $1/2\pi$ times the square root of the constant of proportionality,

i.e. $\qquad f_n = \frac{1}{2\pi} \sqrt{\frac{\lambda}{m}}$

This is the *undamped natural frequency* of the system.

In this formula, λ is the spring rate in newtons per metre, and m is the mass in kilograms.

DAMPED NATURAL FREQUENCY

In real life, of course, the system continually loses energy to its surroundings, due to the damping effect of air resistance and hysteresis, so that the amplitude diminishes to zero in the manner shown in Fig. 10.10. The damping ratio, ζ, can be calculated from

Fig. 10.10 Measurement of the amplitudes of successive half-cycles

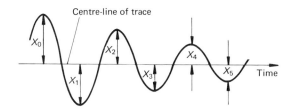

the decrease in amplitude of successive half cycles of free oscillation, by means of the following formula:

$$\zeta = \frac{\delta}{\sqrt{\pi^2 + \delta^2}}$$

where $\delta = \ln(X_0/X_1) = \ln(X_1/X_2) = \ln(X_2/X_3) = $ etc.

and X_0, X_1, X_2, etc., are measured as shown in Fig. 10.10.

This formula is particularly useful for calculating the value of ζ when the oscillation is so heavily damped that only one overshoot can be seen. Where the damping is so light that successive half cycles are approximately equal in amplitude, we can wait for n complete cycles before taking our second measurement of amplitude, so that we measure X_0 and X_{2n}. Then, using the fact that $X_0/X_1 = X_1/X_2 = X_2/X_3 = $ etc., we can say that

$$\frac{X_0}{X_{2n}} = \frac{X_0}{X_1} \times \frac{X_1}{X_2} \times \frac{X_2}{X_3} \times \ldots \times \frac{X_{2n-1}}{X_{2n}}$$

$$- \left(\frac{X_0}{X_1}\right)^{2n}$$

$\therefore$ $$\frac{X_0}{X_1} = \left(\frac{X_0}{X_{2n}}\right)^{1/2n}$$

and $$\delta = \ln\left[\left(\frac{X_0}{X_{2n}}\right)^{1/2n}\right]$$

$\therefore$ $$\delta = \frac{1}{2n} \ln \frac{X_0}{X_{2n}}$$

Besides causing successive half-cycles of oscillation to get smaller, damping also slows down the oscillation, so that the damped natural frequency is less than the undamped. The relationship between the two frequencies is

Fig. 10.11 Graph
showing the effect of
damping on the natural
frequency of a
spring–mass system

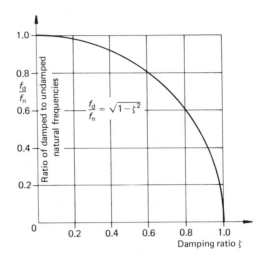

$$\text{Damped natural freq.} = \sqrt{1-\varsigma^2} \times \text{Undamped natural freq.}$$

$$= \sqrt{1-\varsigma^2} \times f_n$$

The effect of damping ratio ς on the frequency of free oscillation, and on the ratio of successive half-cycle amplitudes is shown in the graphs of Figs. 10.11 and 10.12. The ordinates of the graphs are given in the following table:

ς	0	0.001	0.01	0.1	0.2	0.3	0.4	0.6	0.8	1
$\sqrt{1-\varsigma^2}$	1	1.000	1.000	0.995	0.980	0.954	0.917	0.800	0.600	0
X_{n+1}/X_n	1	0.997	0.969	0.729	0.527	0.372	0.254	0.095	0.015	0

Fig. 10.12 Graph
showing the effect of
damping on the decay
of a damped oscillation

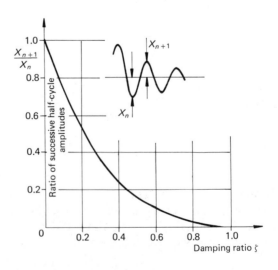

X_{n+1}/X_n is calculated from the formula

$$\frac{X_{n+1}}{X_n} = \exp\left(\frac{-\zeta\pi}{(1-\zeta^2)^{1/2}}\right)$$

which can be deduced from the formula

$$\zeta = \frac{\delta}{\sqrt{\pi^2 + \delta^2}}$$

on p. 205.

The graph of Fig. 10.11 is actually the quadrant of a circle of radius 1.0, since the equation of a circle is $y = \sqrt{r^2 - x^2}$. The graphs and table are worth looking at, as they show us that even quite small values of damping ratio ζ can cause a noticeable reduction of *amplitude* with every half-cycle of oscillation, while the *frequency* is hardly affected at all for damping ratios up to about 0.1.

SELF-TEST QUESTION 17 (Solution on p. 242)

When the output of an automatic control system was monitored by means of an ultraviolet recorder, a step input to the control system resulted in the trace shown in Fig. 10.13. The paper speed was 100 mm/s. Determine (a) the damping ratio, and (b) the *un*damped natural frequency of the control system.

Fig. 10.13

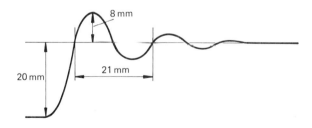

RESPONSE TO A SINUSOIDAL INPUT

Referring again to the damped spring–mass system of Fig. 11.1, suppose we give it an input displacement, x_i, which continuously obeys the equation $x_i = X_i \sin \omega t$. Then the graph of input plotted against time is a *sine wave* or *sinusoid* as shown in Fig. 10.14.

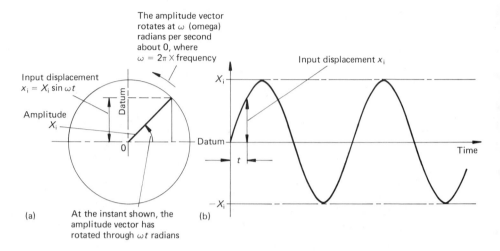

Fig. 10.14 (a) Generation of a sinusoidal input for the system shown in Fig. 10.1. (b) Graph of the sinusoidal input

The output of the system will be another sinusoid *with exactly the same frequency as the input*. The output amplitude may be larger or smaller than the input amplitude, and its motion will lag behind that of the input with a *phase* lag of between 0 and 180° (i.e. between 0 and one half of a cycle of oscillation). Fig. 7.6 on p. 160 shows how the amplitude ratio (i.e. the gain) and the phase lag are calculated.

The amplitude ratio and phase lag of a system depend on the damping ratio, ζ, and on the ratio of input frequency to undamped natural frequency (f/f_n). Figs. 10.15 and 10.16 show, for systems with various values of ζ, how the amplitude and phase lag vary when the input frequency is altered. These curves apply to *second-order systems* — that is, to damped spring–mass systems and to any other systems (e.g. simple automatic control systems) whose output/input relationship can be expressed by a differential equation of the second order. Most automatic control systems are more complicated than this, and their frequency response curves have to be built up from the frequency response curves of their components.

The main points to remember about Figs. 10.15 and 10.16 are:

(a) If f/f_n is very small, the amplitude ratio is approximately 1.0 and the phase lag is little more than zero, for any value of ζ.

(b) As f/f_n becomes very large, the amplitude ratio approaches zero, and the phase lag approaches 180°, for any value of ζ.

(c) When $f/f_n = 1.0$ the phase lag is 90°, for any value of ζ.

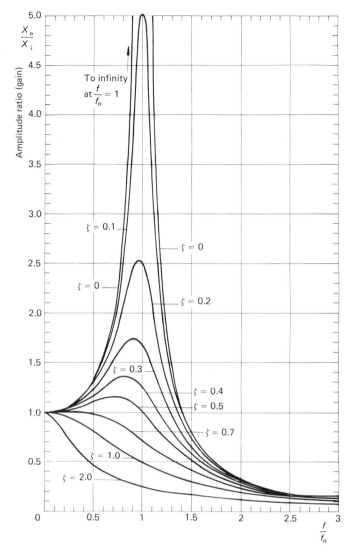

Fig. 10.15 Steady state response of a second-order system, with undamped natural frequency f_n, to a simple harmonic input of frequency f, for various values of damping ratio ζ

(d) As f/f_n increases from zero, the amplitude ratio diminishes for systems with ζ greater than about 0.7, but for systems with ζ less than 0.7, the amplitude ratio increases, reaching a peak at some value of f/f_n rather less than 1.0, then diminishing at higher values of f/f_n, in accordance with point (b) above.

(e) If ζ is very small, there is a sudden increase of phase lag from approximately zero to approximately 180°, as f/f_n passes through 1.0 while increasing. The larger the value of ζ, the more gradual the phase change at this point.

Fig. 10.16 Phase lag (angle by which the output lags behind the input) of a second-order system with undamped natural frequency f_n, subjected to a simple harmonic input of frequency f

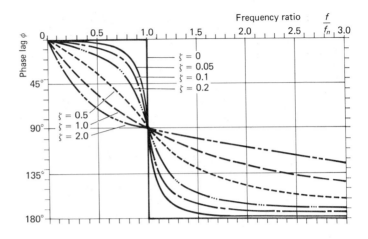

RESONANCE

Fig. 10.15 shows that if the damping ratio of a system is very small, and the system is excited with a sinusoidal input at a frequency close to the undamped natural frequency, the output amplitude may be many times the input amplitude. This effect is called *resonance*. It can cause serious problems in engineering structures, which are apt to act as spring–mass systems in which the spring is the springiness of the structure, and the mass is the mass of the material in the structure. A simple structure such as a propeller shaft or a steel floor joist has very little damping, and if it is acted on by a regularly repeated force, such as an unbalanced centrifugal force rotating at a speed close to the structure's natural frequency, the amplitude ratio can often reach 50 or more. That is, the deflection of the structure may be fifty times the deflection which would be obtained if the cyclic force were applied as a static load. This, in itself, may be serious enough, but a more serious problem may be fatigue of the material, caused by repetition of the deflection many millions of times.

To understand what happens in a system at resonance, let us assume that the damping is small enough to be negligible, so that the block diagram of Fig. 10.9 applies.

Going round the loop from A to D in Fig. 10.9, the signal has changed from $-x_0$ to x_0. That is, there has been a phase change of 180°. The effect of the 180° phase change is to turn the negative feedback into positive feedback.

If we feed energy into the undamped system by applying a sinusoidal input with the same frequency as the resonant frequency, the amplitude of oscillation will increase until something breaks. If there is some damping (as there must be in any real system), the amplitude will be limited in accordance with the curves of Fig. 10.15. It can be shown mathematically that peak amplitude ratio occurs at an input frequency of $\sqrt{1-2\zeta^2}$ times the undamped natural frequency. If the damping ratio, ζ, is small, the differences between this frequency, the damped natural frequency and the undamped natural frequency are negligible. Therefore we can say that:

> *if the damping ratio is small, resonance occurs when the applied frequency equals the natural frequency of the system, and there is positive feedback.*

In practice, the sinusoidal input usually gets applied as a sinusoidally varying force, $F \sin \omega t$, applied to the mass. The effect on the system is just the same as if input had been the sinusoidal displacement of the end of the spring, which we started with in Fig. 10.1. The amplitude ratio and phase lag curves, Figs. 10.15 and 10.16 are unchanged; the only difference is that the amplitude ratio X_0/X_i of Fig. 10.15 becomes the *dynamic magnification factor* or *Q-factor*, X_0/X_s, where X_s is the static deflection due to a force F, i.e. the deflection of the mass which would occur if the amplitude of the sinusoidal force were *steadily* applied.

SELF-TEST QUESTION 18 (Solution on p. 243)

(a) A load carried by a steel beam at mid-span was suddenly released, causing the beam to vibrate. The initial amplitude was 20 mm. After 5.1 seconds, 20 complete cycles of vibration had been counted, and the amplitude had decreased to 2.5 mm. Determine:
   (i)   the damped natural frequency;
   (ii) the damping ratio;
   (iii) the undamped natural frequency;
   (iv) the frequency of excitation at which the *Q*-factor would be a maximum.

(b) In a static test on the beam, a concentrated load of 200 N, applied at mid-span, produced a deflection of 17.5 mm. Determine:
   (i) the spring rate, and
   (ii) the effective mass of the beam.

TRANSIENT AND STEADY-STATE RESPONSES

The curves of Fig. 10.4 illustrate various kinds of response to a step input. Eventually, in all cases, the system response settles down to a steady value. The curves are therefore *transient* responses (the word literally means *passing*), while the final level to which they decay is the *steady-state* response.

In theory, there will be the same two stages of response to a sinusoidal input. That is, if we could suddenly switch on the sinusoidal exciting force, the appropriate transient curve of Fig. 10.4 would be added to the sinusoidal oscillation. Eventually, however, the transient would die away and only the pure sinusoidal response would remain. Fig. 10.17 illustrates this.

Fig. 10.17 Theoretical response to a suddenly applied sinusoidal input

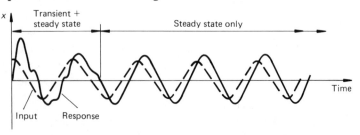

In real life, the sinusoidal force causing the oscillation comes from some mechanical object such as an out-of-balance motor. It would be impossible for such a device to leap instantly from rest to its steady running speed, and so the combination of transient and steady-state responses shown in Fig. 10.17 is unlikely ever to occur in practice, and need not concern us.

REDUCING THE AMPLITUDE OF VIBRATION

The curves of Fig. 10.15 show that if we have a lightly damped steel structure which is liable to be excited into vibration by a sinusoidal force, we must try to prevent the frequency of excitation from coinciding with the natural frequency of the structure.

If we cannot alter the frequency of the exciting force, we might be able to alter the undamped natural frequency, f_n, of the structure.

Suppose we try to increase it, so that the frequency ratio f/f_n is much smaller than 1.0. The formula

$$f_n = \frac{1}{2\pi} \sqrt{\frac{\lambda}{m}}$$

shows us that to increase f_n we must increase λ or reduce m. Unfortunately increasing the stiffness, λ, of the structure is liable also to increase the mass, m, so if we are not careful, we could add a lot of extra material and find that f_n is still much the same as it was. However, by intelligent design (for example, using beams of deeper cross-section) it is often possible to increase λ without increasing m very much.

Alternatively, we could try to make the frequency ratio f/f_n much larger than 1.0. This implies that the undamped natural frequency must be reduced, and hence that λ must be made smaller or m be made larger. The result is liable to be a heavier, more wobbly structure. However, it is a solution often adopted where the amplitude of vibration must be kept to the absolute minimum — in equipment carried on anti-vibration mountings, such as the turntable unit of a record player, or the drum of a spin-dryer, for instance.

One difficulty of working with a frequency of exciting force, f, which is normally much larger than f_n is the 'starting up' problem. If the exciting force is caused by some out-of-balance rotating machine, such as a motor, it does not instantly leap to frequency f when started up, but takes time to accelerate to that frequency. As it does so, it must pass through the region around $f/f_n = 1.0$ in which the amplitude ratio, or Q-factor, is very large, so the mass could oscillate violently. However, feeding enough energy into a spring–mass system to produce large amplitudes of oscillation takes a little time, so, provided the motor, or whatever it is, accelerates rapidly through the high amplitude-ratio region, the peak amplitude attained in this region should be tolerable.

The above remarks apply to (virtually) undamped spring–mass systems, excited at one particular frequency. What if the excitation may take any and every frequency, or what if, even with careful design, the amplitude of vibration would be unacceptably high? A typical example is seen in the suspension of a motor vehicle, where a spring–mass system, the road-wheel and spring, may be excited at any frequency, depending on the speed of the vehicle and the undulations of the road, and yet the wheel must not bounce so badly that the tyre loses contact with the road. The solution in the case of the motor vehicle is to fit a dashpot (i.e. a shock absorber) between the wheel and the vehicle body, to absorb kinetic energy from the spring–mass system and dissipate it as heat.

So if there is no other way to cut down the amplitude of oscillation in a spring–mass system, we can extract energy from it by damping. We can do this either by using a dashpot, or by using high-hysteresis resilient materials for the 'spring'. Alternatively we may de-tune the structure by adding a tuned vibration absorber. These methods are explained below.

THE DASHPOT

The 'classic' form of dashpot consists of a piston and cylinder containing oil. Holes are drilled through the piston to allow oil to pass with difficulty from one side of the piston to the other — alternatively the same effect may be obtained by making the piston a very loose fit in the cylinder. The result is that if the piston is moved along the axis of the cylinder, the drag of oil exerts an opposing force, proportional to the velocity of the movement. This kind of damping is called *viscous damping*. Fig. 10.18 shows the construction of a dashpot.

Fig. 10.18 Oil-damped
dashpot

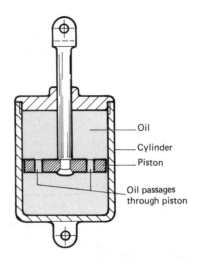

Oil

Cylinder

Piston

Oil passages
through piston

When only a light damping force is necessary, air may be used instead of oil. Air from the atmosphere automatically fills any space, and costs nothing. It enables lighter, simpler dashpots to be designed. Fig. 10.19 shows an air dashpot designed to damp out the oscillations of the pointer of a moving-iron ammeter. (A moving-*coil* meter is automatically damped by eddy currents generated in the aluminium coil former as it moves through the magnetic field — see p. 9.) In Fig. 10.19, any rotation of the moving-iron spindle makes the air trapped on one side of the damping vane escape to the other side through the narrow gap between the vane and the semicircular box in which it is enclosed.

Fig. 10.19 Air-damped
dashpot of a moving-
iron ammeter

Scale of current

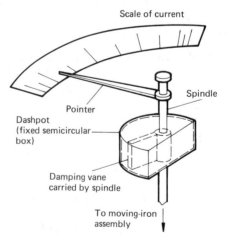

Spindle

Pointer

Dashpot
(fixed semicircular
box)

Damping vane
carried by spindle

To moving-iron
assembly

DAMPING BY HIGH-HYSTERESIS RESILIENT MATERIALS

If a steel spring is loaded and unloaded in alternate directions, it returns almost exactly to its original length — the discrepancy is so small that it needs very careful measurement to detect it. The difference in length after loading in alternate directions is a *hysteresis* effect (see pp. 15–16). Other common structural materials, notably cast iron and aluminium alloy, have more hysteresis than steel, and hence have more internal damping than steel has. We can hear the difference if we drop a bar of each material on to a concrete floor. The steel will ring like a bell, the cast iron has a duller sound, and the aluminium alloy hardly resonates at all — an indication of their relative vibration-damping characteristics. So cast iron is used instead of steel for lathe beds, not only because it can be cast in complex shapes, but also because it is better at damping out machine tool chatter. And aluminium alloy is used for aircraft structures, not only because it is as light as steel for the same strength, but also because it is better for damping out 'flutter'-induced vibrations.

Rubber and highly polymeric plastics have quite high internal damping, and are therefore very suitable for use as a combined spring and damping agent, when we want to design a compact damped spring–mass system. A commonly used high polymeric plastic is softened polyvinylchloride (PVC).

These materials undergo both elastic and plastic deformation when under load. The elastic deformation disappears immediately the load is removed, but the plastic deformation disappears only after some time lag.

A typical use for such materials is in anti-vibration mountings. A motor-car engine, for instance, is mounted on rubber blocks, to minimise the transmission of vibration from the engine to the vehicle body. The blocks should ideally be arranged to deflect in shear as shown in Fig. 10.20(a), because a block loaded in compression, as shown in Fig. 10.20(b), gets fatter as it compresses, and this increases the spring rate — the block tends to 'go solid' under compression.

Fig. 10.20 Deflection of a rubber block: (a) in shear, (b) in compression

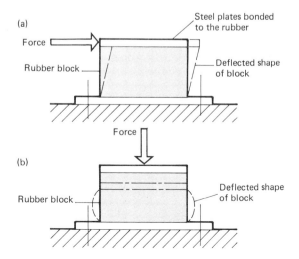

It is a little difficult to arrange engine mounting blocks to be in *pure* shear, because an engine vibrates vertically and horizonally, as well as torsionally. The best compromise is to set the blocks at 45° to the vertical, as shown in Fig. 10.21.

Fig. 10.21 Anti-vibration mounting for a motor vehicle engine

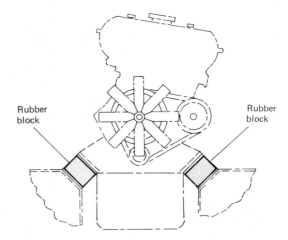

Rubber block

Rubber block

A car engine which is only just 'ticking over' bounces violently on its rubber blocks, because the alternating inertia forces have about the same frequency as the natural frequency of the spring–mass system (the rubber being the spring and the engine the mass). However, the oscillations die out instantly when the engine stops, due to the heavy damping effect of the rubber. When the engine is running at high speed, however, its amplitude of oscillation is very small indeed, because the inertia forces have a much higher frequency and thus f/f_n (see Fig. 10.15) is very much greater than 1.0.

The advantages of using rubber and plastic materials as damped springs are cheapness, compactness, simplicity of design, and the fact that the ratio of stiffness to damping can be varied by varying the composition and treatment of the material during manufacture. A disadvantage is that the properties of the material may change over a long period of time.

THE TUNED VIBRATION ABSORBER

This is a way of reducing the vibration of a spring–mass system by attaching another spring–mass system to it, so that the vibrations of the added system approximately cancel out those of the original system.

Figs. 10.22 and 10.23 show the principle as it would apply to a machine tool if we wished to reduce its amplitude of vibration by this method. Fig. 10.22(a) shows that we can think of the machine tool as a mass (the mass of the cutting tool and its immediate

Fig. 10.22 (a) Damped spring–mass equivalent of a machine tool. (b) Typical frequency response of the system

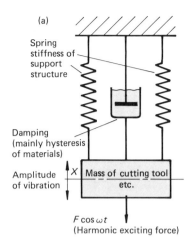

(a)

Spring stiffness of support structure

Damping (mainly hysteresis of materials)

Amplitude of vibration

X Mass of cutting tool etc.

$F\cos\omega t$ (Harmonic exciting force)

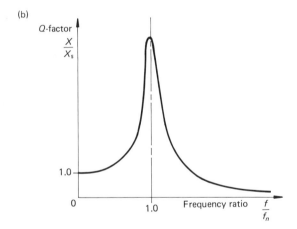

(b)

Q-factor

$\dfrac{X}{X_s}$

1.0

0

1.0   Frequency ratio $\dfrac{f}{f_n}$

support), a spring (the stiffness of the machine frame), and some damping (the hysteresis of the material of the machine frame). This kind of damping could only be very light, and so the frequency response curve would have a very large Q-factor at the natural frequency of the system, as shown in Fig. 10.22(b) which represents one curve from the family of curves originally shown in Fig. 10.15.

Fig. 10.23(a) represents the system with a tuned vibration absorber added. If the additional spring and mass have negligible damping, it is possible to calculate their values so that near the original resonant frequency, the Q-factor is now virtually zero, but there would be large Q-factor peaks on either side of this frequency as shown by the unbroken curve of Fig. 10.23(b). It is better to provide fairly strong damping for the added mass, so that the Q-factor remains approximately constant but does not greatly exceed 1.0, as shown by the dotted curve of Fig. 10.23(b).

Fig. 10.23 (a) Equivalent
system of Fig. 10.22(a)
with tuned vibration
absorber added. (b)
Typical effects of the
vibration absorber on
the curve of Fig. 10.22(b)

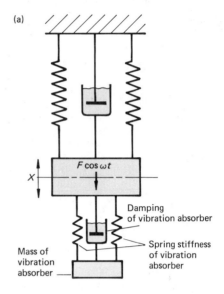

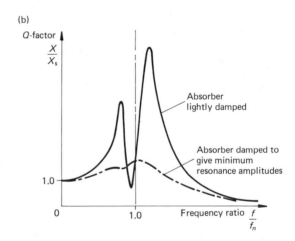

THE PREVENTION OF MACHINE TOOL CHATTER

During machining processes such as turning or milling, the
machine behaves as a spring–mass system in which there is very
little damping.

If the feed and depth of cut are kept small, the chips of metal
produced by the cutting process are of uniform thickness, and
the vibration of the spring–mass system is negligible, so the workpiece is produced with a smooth surface finish. However, for
economical production, it is often necessary to operate with greater
feed rates and depths of cut than the ideal. Under these conditions,
the mass tends to vibrate at the natural frequency of the
spring–mass system, causing variations in depth of cut, and undulations (waves) in the machined surface, as shown in Fig. 10.24. The

next time the cutting edge ploughs through this undulating surface (e.g. on the next revolution of an object being turned) the undulations made in the previous pass cause variations in the cutting force at the same frequency (the natural frequency of vibration of the system) and these keep the vibration going. This effect, *machine tool chatter*, is thus a self-excited vibration, which produces ripples and furrows in machined surfaces intended to be smooth, and which reduces the life of the tool and the machine. To suppress it, the following methods may be used.

Fig. 10.24 Wavy surface caused by vibration (chatter) between tool and workpiece

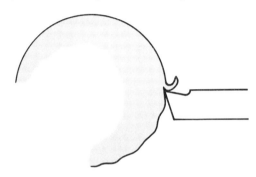

VISCOUS-DAMPING A LATHE SPINDLE

The most flexible part of a lathe cutting system is probably the spindle which carries the chuck. It forms the spring of a spring–mass system, the mass being the chuck and workpiece. If it vibrates as a beam, the chuck wobbles and the result is tool chatter.

Fig. 10.25 shows a method of damping this kind of vibration by applying the dashpot principle. The 'dashpot' in this case is a bush which is slightly larger than the spindle, and which is fixed inside an oil-filled chamber. Any radial movement of the spindle forces oil through ventholes which connect circular grooves on the inside of the bush to the oil-filled chamber on the outside. Thus vibrations which cause bending of the spindle are quickly damped out, by using up their energy to force oil through the ventholes.

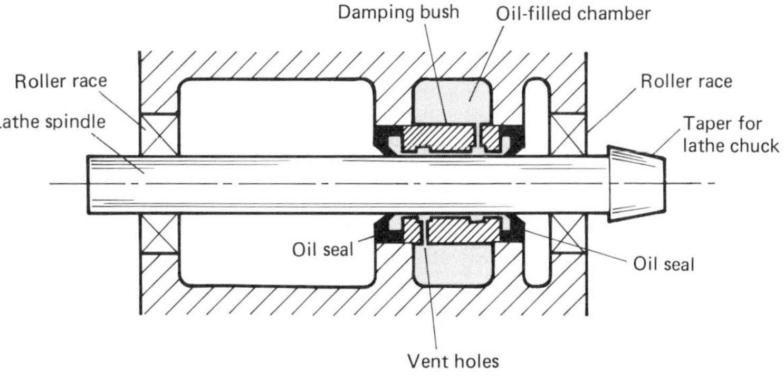

Fig. 10.25 Viscous damper to suppress the bending vibration of a lathe spindle

Damping bush   Oil-filled chamber

Roller race   Roller race

Lathe spindle   Taper for lathe chuck

Oil seal   Oil seal

Vent holes

DAMPING A BORING BAR

A long overhung boring bar forms a spring–mass system in which the bar acts as both the spring and the mass. Any device for damping its vibrations has to be carried inside the bar, because the bar must operate inside the workpiece.

Fig. 10.26 shows how the principle of the tuned vibration absorber (illustrated in Fig. 10.23) may be applied to a boring bar. For best effect, the additional vibratory system ought to be carried in an extension of the boring bar beyond the cutting tool (i.e. to the right of it in Fig. 10.26), but this is not usually possible if the bar has to be used to bore right to the end of blind holes.

Fig. 10.26 Tuned vibration absorber for a boring bar, with viscous damping

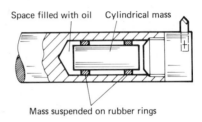

Space filled with oil   Cylindrical mass

Mass suspended on rubber rings

The additional vibratory system consists of a cylinder of heavy metal (cemented carbides, tungsten or lead) supported in rubber rings, which act as both the spring and the damper. Additional damping may be provided by filling the hole in the bar with oil.

A simpler device is the shock-damper shown in Fig. 10.27. This uses the fact that any impact between two bodies dissipates kinetic energy as heat. In this case the impacts occur between a cylinder of heavy metal and the walls of the clearance hole in the boring bar in which it lies.

Fig. 10.27 Shock-damper for a boring bar

Loose cylindrical mass

Clearance allows mass to absorb energy by impact

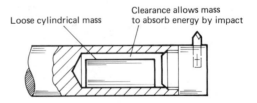

The effectiveness of the device may be increased as shown in Fig. 10.28, by making the loose mass in the form of a number of short cylindrical slugs sandwiched between discs which fit the hole in the boring bar. A compression spring keeps the discs and slugs pressed against each other. This arrangement extracts energy from the vibration both by the impact of the slugs on the walls of the hole, and by friction between the slugs and the discs.

Fig. 10.28 A boring bar with combined shock-damping and dry friction damping

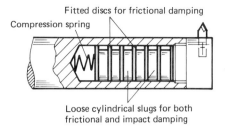

Fitted discs for frictional damping

Compression spring

Loose cylindrical slugs for both frictional and impact damping

It should be noted that the clearance between the impact masses and the walls of the hole has had to be greatly exaggerated in Figs. 10.27 and 10.28. Only a very small amplitude of vibration is necessary for tool chatter to become objectionable, and before this amplitude is reached, the mass should be striking alternate sides of the hole. So, for the shock damper to be effective, the clearance between the mass and the hole should be very small indeed.

DAMPING THE OVERARM OF A HORIZONTAL MILLING MACHINE

A horizontal milling machine may be damped by adding a tuned vibration absorber as shown in Fig. 10.29. This consists of a mass suspended in a U-shaped bracket by discs of rubber or softened PVC bonded to both the mass and the bracket. Inertia forces on the mass cause the rubber or plastic to deflect in shear; thus it acts as both the spring and the damping agent of the additional vibratory system. This system forms a self-contained unit, which may be bolted to the overarm.

Fig. 10.29 Tuned vibration absorber for the overarm of a horizontal milling machine

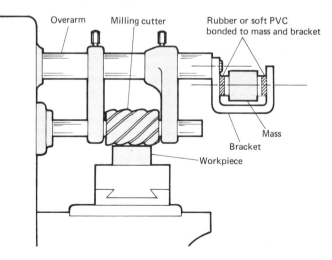

Overarm   Milling cutter   Rubber or soft PVC bonded to mass and bracket

Mass

Bracket

Workpiece

The values of mass, spring stiffness, and damping ratio for a tuned vibration absorber would be calculated to suit the particular machine to which it was to be applied, and would require final adjustment after it had been attached to the machine, in order to obtain the most effective suppression of vibration possible.

SUMMARY OF VIBRATION FORMULAE

$$\text{Damping ratio } \zeta = \frac{\text{Actual damping}}{\text{Critical damping}}$$

$$\text{Undamped natural frequency } f_n = \frac{1}{2\pi}\sqrt{\frac{\lambda}{m}}$$

Damping ratio from measurement of a decaying oscillation:

$$\zeta = \frac{\delta}{\sqrt{\pi^2 + \delta^2}}$$

where $\delta = \ln\dfrac{X_0}{X_1} = \ln\dfrac{X_1}{X_2} = \ln\dfrac{X_2}{X_3} = \text{etc.}$

Also $\delta = \dfrac{1}{2n}\ln\dfrac{X_0}{X_{2n}}$ where n is number of complete cycles

Effect of damping

$$\text{Damped natural frequency } f_d = \sqrt{1-\zeta^2} \times f_n$$

$$\text{Period} = \frac{1}{\text{Frequency}}$$

$\therefore$ $\text{Period of damped vibration} = \dfrac{1}{\sqrt{1-\zeta^2}} \times \text{Period of undamped.}$

Peak amplitude ratio occurs at $\sqrt{1-2\zeta^2} \times f_n$

EXERCISES ON CHAPTER 10

1   (a) By means of sketched graphs on the same axes, show how the response of a spring–mass system to a step input varies when the system is:
(i)   underdamped;
(ii)   overdamped;
(iii) critically damped.

(b) Explain why some instruments are designed to have a damping ratio of about 0.7.

2   Why do undamped spring–mass systems vibrate freely when disturbed?

3   When a damped spring–mass system was displaced 15 mm from the equilibrium position and then released, the first overshoot was 5 mm on the other side of the equilibrium position, and two complete cycles of oscillation were found to occupy 2.4 seconds. Calculate:

(a) the damping ratio;
(b) the damped natural frequency; and
(c) the undamped natural frequency.

4   A measuring instrument is found to behave as a damped spring–mass
    system when subjected to a step input. Its first overshoot is 30%
    of the steady-state change, and its periodic time of oscillation is
    0.8 seconds. Choose, from the following, the time–range within
    which the settling time will lie, if the allowable tolerance limits are
    ± 3% of the steady-state value:

    (a)   (b)   (c)   (d)
    0.4–0.6 s   0.6–0.8 s   0.8–1.0 s   1.0–1.2 s

    (e)   (f)   (g)
    1.2–1.4 s   1.4–1.6 s   1.6–1.8 s

5   Determine:
    (a) the damping ratio;
    (b) the damped natural frequency; and
    (c) the undamped natural frequency of the instrument in
         Question 4.

6   The specification of the instrument in Question 4 is to be relaxed
    so that its response to a step input can now be within ± 3% of the
    steady-state value in not more than three complete cycles of
    oscillation. What is the maximum percentage which can now be
    allowed for the first overshoot?

7   A mass of 15 kg suspended from a spring performs vertical damped
    oscillations. The amplitude of the vibration is found to reduce to
    25% of the original value after 20 complete oscillations. 100
    oscillations are completed in a time of 42 seconds. Calculate:

    (a) the damping ratio and
    (b) the stiffness of the spring.

8   A system consists of a spring of stiffness 4 kN/m and a mass of
    5 kg. When it is vibrating freely, the amplitude of vibration decays
    1% for each full cycle. Calculate:

    (a) the damping ratio and
    (b) the frequency of the vibration.

9   Sketch graphs of amplitude ratio and phase lag for damping ratios
    of 0.1, 0.5, 1.0 and 2.0 to show how the motion of spring–mass
    systems varies as the forcing frequency is varied from 0 to 3 times
    the undamped natural frequency.

10 The phase lag between the output and input of a spring–mass
    system subjected to a forced oscillation may vary between $0°$ and
    $180°$. What can we deduce if the phase lag is:
    (a) less than $90°$?
    (b) equal to $90°$?
    (c) greater than $90°$?

11 Define 'resonance'. Why is it not strictly true to say that resonance occurs when the forcing frequency is equal to the undamped natural frequency? Under what circumstances is the statement approximately true?

Questions 12 to 15 give information obtained from tests to determine the dynamic characteristics of some second-order systems. Each question relates to a different system. Use Figs. 10.15 and 10.16 to estimate, from the information given in the question, (a) the damping ratio, and hence determine (b) the undamped natural frequency, (c) the amplitude ratio at 300 Hz and (d) the phase lag at 300 Hz.

12 Peak value of amplitude ratio was found to be 2.54 at a frequency of 100 Hz.

13 Peak value of amplitude ratio was found to be 1.5 at a frequency of 150 Hz.

14 A frequency of 200 Hz gave a phase lag of 90° with amplitude ratio of 0.5.

15 A frequency of 90 Hz gave a phase lag of 70° with amplitude ratio 0.46.

16 (a) Define the terms 'frequency ratio', 'damping ratio' and 'magnification factor'.
    (b) A mass of 12 kg is supported by an elastic structure of stiffness 3000 N/m. The mass is vibrated by a simple harmonic disturbing force having a constant peak value of 100 N. The damping ratio is 0.4. Determine:
    (i)   the undamped natural frequency;
    (ii) the resonant frequency;
    (iii) the amplitude at the resonant frequency (estimate from Fig. 10.15).

17 A machine with a mass of 30 kg rests on a damped 'springy' support which has a stiffness of 100 kN/m. When the machine runs at its normal speed of 440 rev/min, it is acted on by a simple harmonic disturbing force at the same frequency as its running speed, with a peak value of 150 N. As a result, the engine vibrates on its supports with an amplitude of 2.5 mm. Determine:
    (a) the frequency ratio and
    (b) the Q-factor;
    and hence estimate, using the curves of Fig. 10.15
    (c) the damping ratio of the system.

18 It is necessary to reduce the amplitude of the steady-state vibration of the machine in Question 17 to 1 mm. If this is to be done by adding mass to the machine, leaving everything else unchanged:

    (a) use the curves of Fig. 10.15 to estimate the frequency ratio which would be necessary.

    (b) hence determine the total mass of machine plus added ballast.

19 (a) What is the effect of exciting a lightly damped spring–mass system at a forcing frequency close to the resonant frequency of the system?

    (b) State two alternative ways in which the designer might try to (i) increase, or (ii) decrease the resonant frequency of the system.

    (c) If he decreases the resonant frequency of the system, what operating difficulty must be guarded against?

    (d) If neither the resonant frequency nor the exciting frequency can be changed, what are the other two means which can be used to reduce the amplitude of vibration of the system?

20 Sketch and describe the operating principle of a dashpot.

21 How is damping effected:

    (a) in a moving coil meter,

    (b) in a moving iron meter;

    (c) on the spring suspension of a motor car;

    (d) on the engine mounting of a motor car?

22 What materials are used for hysteresis damping?
Discuss the advantages and disadvantages of using this kind of damping instead of viscous damping.

23 With the aid of a diagram of the system and sketches of frequency response curves explain the principle of the tuned vibration absorber.

24 (a) What is machine tool chatter, and what are its effects?

    (b) Explain with the aid of sketches how machine tool chatter can be reduced by:

        (i)   using an impact damper in a boring bar;

        (ii) viscous-damping the spindle of a lathe;

        (iii) attaching an additional vibratory system to the overarm of a horizontal milling machine.

Miscellaneous Exercises

1   The following tables list some transducers, some measurement systems and some display devices. Complete the tables by single-word answers or ticks, as appropriate.

Transducers	Basic units of input	Basic units of output	Electrical power supply needed?	
			Yes	No
Thermistor				
Potentiometer				
Piezoelectric crystal				
Photoresistive cell				
Tachogenerator (transmitter only)				
Bourdon tube (tube only)				
Proving ring (ring only)				
Manometer				

Measurement systems	Basic units of input	Form of signal entering display device	Display device	Electrical power supply needed?		Physical contact with measured object needed?	
				Yes	No	Yes	No
Dial test indicator							
Linear variable differential transformer							
Eddy current tachometer							
60-toothed wheel and magnetic transducer							
Strain-gauged load cell							

Display devices	Basic units of input	Form of display (tick one column)				Max. frequency (tick one column)				
		Instantaneous value	Transient input/time	Permanent input/time	Graph of two inputs	0–10 Hz	10–10^3 Hz	10^3–10^5 Hz	10^5–10^7 Hz	10^7–10^9 Hz
Moving coil meter										
Ultraviolet recorder										
X–Y plotter										
Y–t plotter										
Cathode-ray oscilloscope (tick two kinds of display)										
Pulse counter										

2 For each of the transducers listed below, select three of the statements (*a*) to (*g*) which apply to it. Your answer for each of the five transducers should consist of three letters: one will be (*a*) or (*b*); one will be (*c*), (*d*) or (*e*); and one will be (*f*) or (*g*).

Transducers

(i)   variable capacitor

(ii) strain gauge

(iii) wire-wound potentiometer

(iv) piezoelectric crystal

(v)   linear variable differential transformer.

Statements

(a) There must be contact between fixed and moving parts of the transducer.

(b) There may not be contact between fixed and moving parts of of the transducer.

(c) Requires a.c. excitation.

(d) Needs no excitation.

(e) Can use d.c. excitation.

(f) Has infinite resolution.

(g) Has finite resolution.

3 (a) Show by means of a diagram the arrangement of the essential parts of a moving-coil meter and explain how it works.

(b) How is it adapted to serve as:
   (i)   an ammeter;
   (ii) a voltmeter;
   (iii) an ohm-meter;
   (iv) a mirror galvanometer?

(c) Which of these four adaptations would you use to display the output of:
   (i) a potentiometer type of displacement transducer;
   (ii) a strain-gauge bridge working on the null balance method, in which maximum sensitivity is required?

4 You are required to propose instrumentation for an engine test bed for the routine testing of petrol engines coming off a production line. The following measurements are to be continuously displayed in a windowless control room at some distance from the test bed:

(a) engine torque output;

(b) crankshaft rotational speed;

(c) fuel flow rate;

(d) cooling water temperature;

(e) exhaust gas temperature at outlet from manifold.

Decide on a means of measuring each of the above quantities, and in each case explain, with diagram and brief description, the operating principle of the measurement system you have adopted.

5 In each of the cases (a) to (f) below, specify a suitable item of equipment, and briefly justify your choice by explaining its particular advantage for the purpose required:

(a) a display device for a high frequency waveform;

(b) a transducer for measuring the level of liquid in a tank, the measurement to be displayed at some distance from the tank;

(c) a recorder for recording several inputs simultaneously, with frequencies ranging from zero to a few kilohertz;

(d) a transducer for measuring linear displacements in the range 0 to 10 mm with a resolution of 0.1 mm; there must be no frictional resistance between the transducer and the measured object;

(e) a recorder to give an immediate graph of two electrical signals with frequency components up to 10 Hz;

(f) a system to give an a.c. signal carrying information on the degree of angular misalignment of two shafts.

Solutions to Self-test Questions

QUESTION 1, p. 32

(a) Change of resistance,

$$R = 120.42 - 120.27$$

$$= 0.15\,\Omega$$

$$\frac{\delta R}{R} = k\epsilon$$

$$\therefore \quad \frac{0.15}{120.27} = 2.1\epsilon$$

$$\therefore \quad \epsilon = \frac{0.15}{120.27 \times 2.1}$$

$$= 0.000\,594$$

or   594 microstrain

(b) $\dfrac{\text{Stress}}{\text{Strain}} = 200\,\text{GN/m}^2$

$\therefore$ $\text{Stress} = 0.000\,594 \times 200 = 0.1188\,\text{GN/m}^2$

$$= 118.8\,\text{MN/m}^2 = 118.8\,\text{N/mm}^2$$

(c) $\dfrac{\text{Load}}{\text{Area}} = \text{Stress}$

$\therefore$ $\dfrac{F}{25 \times 6} = 118.8\,\text{N/mm}^2$

$\therefore$ $F = 118.8 \times 25 \times 6 = 17\,820\,\text{newtons} = 17.82\,\text{kN}$

QUESTION 2, p. 33

$$\frac{\text{Stress}}{\text{Strain}} = \text{Modulus of elasticity}$$

$\therefore$ $\text{Strain} = \dfrac{\text{Stress}}{\text{Modulus}}$

Taking tensile values as positive and compressive as negative gives the compressive stress as $-30\,\text{N/mm}^2$, which is $-30\,\text{MN/m}^2$, which is $-0.030\,\text{GN/m}^2$.

Therefore the strain is $-0.030/200$, which is $-0.000\,15$.

$$\frac{\delta R}{R} = k\epsilon$$

∴ $$\frac{\delta R}{120.27} = 2.1 \times (-0.000\,15)$$

∴ $$\delta R = 120.27 \times 2.1 \times (-0.000\,15)$$

$$= -0.038\,\Omega$$

Therefore the resistance of the strain gauge is now

$$120.27 - 0.038\,\Omega = 120.232\,\Omega$$

QUESTION 3, p. 37

(a) Fig. S.1 is the required calibration graph. It demonstrates the typical exponential curve characteristic of a thermistor.

Fig. S.1

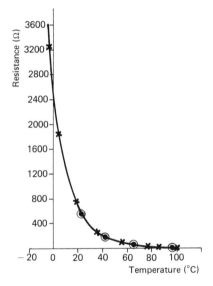

(b) The general law of a thermistor is $R = A\,e^{B/T}$.

Taking logarithms to the base e of each side of the equation we have

$$\ln R = \ln (A\,e^{B/T})*$$

∴ $$\ln R = \ln (e^{B/T}) + \ln A$$

∴ $$\ln R = \frac{B}{T} + \ln A \qquad\qquad\qquad\qquad [1]$$

This is of the form $y = ax + b$ where $\ln R$ corresponds to y and $1/T$ corresponds to x, so if we plot $\ln R$ against $1/T$ we should get a straight line from which the constants B and $\ln A$ (and hence A) can be determined.

* ln is alternative notation for $\log_e$.

The following table is calculated from the results:

°C	R	T kelvin	$\dfrac{1}{T}$ K^{-1}	$\ln R$
−5	3260	268	0.003 73	8.09
3.5	1831	276.5	0.003 62	7.51
17.5	765	290.5	0.003 44	6.64
36	263	309	0.003 24	5.57
55	97.7	328	0.003 05	4.58
76	36.2	349	0.002 87	3.59
85	24.3	358	0.002 79	3.19
100	13.3	373	0.002 68	2.59
97	14.2	370	0.002 70	2.65
66	55.6	339	0.002 95	4.02
43	171.4	316	0.003 16	5.14
21	581	294	0.003 40	6.36

The values in the two right-hand columns are plotted in Fig. S.2 and a good straight line is obtained.

Fig. S.2

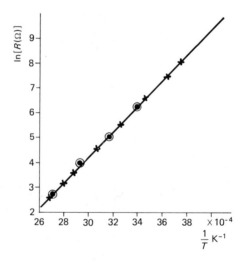

Taking two points on the straight line — say

$$\ln R = 8.09, \quad \frac{1}{T} = 0.003\ 73\ \text{K}^{-1} \quad \text{and}$$

$$\ln R \doteq 3.19, \quad \frac{1}{T} = 0.002\ 79\ \text{K}^{-1}$$

and substituting into equation [1] we have a pair of simultaneous equations:

$$8.09 = 0.003\ 73B + \ln A$$

$$3.19 = 0.002\ 79B + \ln A$$

Then subtracting gives

$$4.9 = 0.000\,094B$$

$$\therefore \qquad B = 5213\,\text{K}$$

Substituting into the first equation

$$8.09 = 0.003\,73 \times 5213 + \ln A$$

$$\therefore \qquad \ln A = -11.35$$

$$\therefore \qquad A = 0.000\,011\,72\,\Omega$$

Therefore the law of the thermistor is $R = 0.000\,011\,72\,e^{5213/T}$.

QUESTION 4, p. 39

The results are plotted in Fig. S.3. The voltage at the instant the step was applied was 6 volts.

$$0.368 \text{ of } 6 = 2.2 \text{ volts}$$

Therefore the time constant (τ) of the system is the time when the curve crosses the level of 2.2 volts. From the graph, this is found to be 5.6 seconds.

The time constant can also be found from the fact that if the output continued to change at its original rate, the output change would be completed in τ seconds. This is because

$$\frac{\text{d}}{\text{d}t}(V\,e^{-t/\tau}) = -\frac{V}{\tau}\,e^{-t/\tau}$$

and when $t = 0$, this becomes $-V/\tau$ because $e^0 = 1$.

So the gradient of the graph at the point when we started counting time is $-V/\tau$, and thus the time it would take for the output voltage of the crystal to fall from V volts to 0 volts at this rate must be τ seconds.

This is illustrated in Fig. S.3. PQ is the tangent to the curve at its starting point, P, and its intersection with the zero volts line

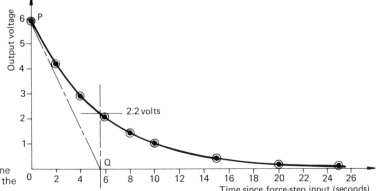

Fig. S.3 Voltage–time graph resulting from the step input

coincides with the time constant time of 5.6 seconds. Thus the tangent method gives us an alternative (but less accurate) way of finding the time constant.

QUESTION 5, p. 40

We need a 'scientific' calculator for this. Taking the first value, 0.01τ, as an example, we say

$$e^{-t/\tau} = e^{-0.01\tau/\tau} = e^{-0.01}$$

Entering 0.01 on the calculator, and pressing the $+/-$ button and then the e^x button we get 0.990 — this means that 99% of original output remains after a time equal to 0.01 of the time constant of the system. Treating the other time intervals in the same way we get the following table:

Elapsed time since the step input (defined as a number multiplying the time constant)	Percentage of original output which remains
0.01τ	99.0%
0.02τ	98.0%
0.05τ	95.1%
0.10τ	90.5%
1τ	36.8%
2τ	13.5%
3τ	5.0%
4τ	1.8%
5τ	0.7%

QUESTION 6, p. 64

The overall gain is $12 \times 25 \times 4 = 1200$

Their gains in dB are:

$$20 \log_{10} 12 = 21.6 \, \text{dB}$$

and
$$20 \log_{10} 25 = 28.0 \, \text{dB}$$

$$20 \log_{10} 4 = 12 \, \text{dB}$$

$$\text{Total} = 61.6 \, \text{dB}$$

For comparison, the overall gain of 1200 is

$$20 \log_{10} 1200 = 61.6 \, \text{dB}$$

Alternatively we could convert the 61.6 dB to a voltage gain by saying

$$20 \log_{10} (\text{voltage gain}) = 61.6$$

$$\therefore \qquad \log_{10}(\text{voltage gain}) = \frac{61.6}{20} = 3.08$$

$$\therefore \qquad \text{Voltage gain} = 10^{3.08}$$

$$= 1202$$

QUESTION 7, p. 70

Fig. S.4 is the graph of the results. Take two points from the best straight line through the plotted points, e.g.

$$\text{Load} = 0, \quad R_B = 800.05\,\Omega \quad \text{and}$$

$$\text{Load} = 20, \quad R_B = 803.0\,\Omega$$

Then,

$$\text{Increase of load} = 20\,\text{kN}$$

$$\text{Area of cross-section} = \frac{\pi}{4} \times 13.9^2\,\text{mm}^2$$

$$= 151.7\,\text{mm}^2$$

$$\therefore \qquad \text{Increase of stress} = \frac{20}{151.7}$$

$$= 0.1318\,\frac{\text{kN}}{\text{mm}^2}$$

$$= 131.8\,\frac{\text{N}}{\text{mm}^2} \quad \text{or} \quad \frac{\text{MN}}{\text{m}^2}$$

Since all four arms of the bridge have nominally equal resistance, we can say

$$\delta R_A = \delta R_B$$

$$= 803.0 - 800.05$$

$$= 2.95\,\Omega$$

Fig. S.4 Graph of results

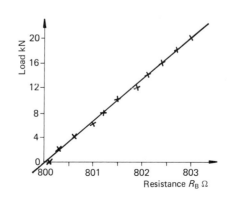

Load kN

Resistance R_B Ω

$$\text{Increase of strain } \epsilon \;=\; \frac{\delta R}{kR}$$

$$=\; \frac{2.95}{2.05 \times 800.1}$$

$$=\; 0.001\ 799$$

$$\text{Modulus of elasticity} \;=\; \frac{\text{Increase of stress}}{\text{Increase of strain}}$$

$$=\; \frac{131.8}{0.001\ 799}\ \frac{\text{MN}}{\text{m}^2}$$

$$=\; 73\ 300\ \frac{\text{MN}}{\text{m}^2}$$

$$=\; 73.3\ \frac{\text{GN}}{\text{m}^2}$$

QUESTION 8, p. 101

A given voltage difference applied to either pair of plates should bend the beam through the same angle, if the speed of the electrons past them is the same. But the further from the screen the bend occurs, the greater the displacement of the spot from the centre of the screen.

Also, in fact, the electrons continue to accelerate until they hit the screen, so they are moving faster through the second pair of plates and hence a given voltage difference bends the beam through a smaller angle.

QUESTION 9, p. 111

Ratio of successive measured speeds	Ratio of successive theoretical speeds
$\dfrac{5050}{3850} = 1.31$	$1 \div \frac{1}{2} = 2$
$\dfrac{3850}{3000} = 1.28$	$\frac{1}{2} \div \frac{1}{3} = 1.5$
$\dfrac{3000}{2525} = 1.19$	$\frac{1}{3} \div \frac{1}{4} = 1.33$
$\dfrac{2525}{2200} = 1.15$	$\frac{1}{4} \div \frac{1}{5} = 1.25$
	$\frac{1}{5} \div \frac{1}{6} = 1.20$
	$\frac{1}{6} \div \frac{1}{7} = 1.167$
	$\frac{1}{7} \div \frac{1}{8} = 1.143$

In this case, it is not quite so obvious which figures in the left-hand column correspond to which figures in the right, because we are using the stroboscope to try to measure a speed far above its maximum flash rate, so that the inevitable slight errors in the readings have a greater tendency to 'blur' the similarities between the ratios in the two columns. However, it looks as if matching 1.31, 1.28, 1.19 and 1.15 to the sequence 1.33, 1.25, 1.20 and 1.167 respectively will cause least discrepancy, so 5050 rev/min corresponds to $\frac{1}{3}$ of the speed, making our estimate of the speed:

$$\frac{3 \times 5050 + 4 \times 3850 + 5 \times 3000 + 6 \times 2525 + 7 \times 2200}{5}$$

$$= \frac{15\,150 + 15\,400 + 15\,000 + 15\,150 + 15\,400}{5}$$

$$= \frac{76\,100}{5}$$

$$= 15\,220 \text{ rev/min}$$

QUESTION 10, p. 124

(a) (i) Gauge pressure $= \dfrac{28.5}{760} \times 101.3 \text{ kPa}$

$$= 3.80 \text{ kPa}$$

(ii) Absolute pressure $= 3.80 + 101.3 = 105.1 \text{ kPa}$

(b) Alternatively,

Gauge pressure $= \rho g h$

$$= 13.6 \times 1000 \times 9.8 \times 0.0285 \frac{\text{kg}}{\text{m}^3} \times \frac{\text{m}}{\text{s}^2} \times \text{m}$$

$$= 3800 \frac{\text{N}}{\text{m}^2}$$

$$= 3.80 \text{ kPa}$$

(c) Corresponding difference in levels of water $= 13.6 \times 28.5$

$$= 388 \text{ mm H}_2\text{O}$$

QUESTION 11, p. 124

(a) $p_1 - p_2 = (13.6 - 1) \times 1000 \times 9.8 \times 0.170 \dfrac{\text{kg}}{\text{m}^3} \times \dfrac{\text{m}}{\text{s}^2} \times \text{m}$

$$= 21\,000 \frac{\text{N}}{\text{m}^2}$$

$$= 21.0 \text{ kPa}$$

(b) $\qquad p_1 - p_2 = (13.6 - 0.8) \times 1000 \times 9.8 \times 0.170$

$$= 21\,300 \, \frac{N}{m^2}$$

$$= 21.3 \, \text{kPa}$$

QUESTION 12, p. 126

(a) The relationship between scale length and vertical height is illustrated by the right-angled triangle shown in Fig. S.6.

Fig. S.6

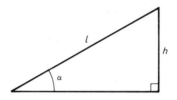

$$\frac{h}{l} = \sin \alpha$$

$\therefore \qquad l = \dfrac{h}{\sin \alpha}$

$$= \frac{40}{\sin 8°} \, \text{mm}$$

$$= 287 \, \text{mm}$$

Therefore the length of one 'millimetre' on the scale is

$$\frac{287}{40} = 7.19 \, \text{mm}$$

(b) (i) $\quad$ Maximum percentage error $= \dfrac{0.5 \, \text{mm}}{10 \, \text{mm}} \times 100\%$

$$= 5\%$$

(ii) $\quad$ Maximum percentage error $= \dfrac{0.5 \, \text{mm}}{10 \times 7.19 \, \text{mm}} \times 100\%$

$$= 0.7\%$$

(c) $\qquad$ Area of internal cross-section in the inclined limb

$$= \frac{\pi}{4} \times 2.5^2 \, \text{mm}^2$$

$$= 4.91 \, \text{mm}^2$$

Volume of liquid contained between the scale readings of '0' and '40'

$$= 287 \times 4.91 = 1411 \, \text{mm}^3$$

Cross-sectional area of enlargement in other branch

$$= \frac{\pi}{4} \times 38^2$$

$$= 1134 \, \text{mm}^2$$

Thus the change in level in the enlarged limb for 'full-scale deflection' is

$$\frac{1411 \, \text{mm}^3}{1134 \, \text{mm}^2} = 1.24 \, \text{mm}$$

This implies that a pressure reading of 40 mm H$_2$O on the scale is actually 41.24 mm.

To put this right, the 'millimetre' divisions on the scale would be shortened to

$$\frac{7.19 \times 40 \, \text{mm}}{41.24} = 6.97 \, \text{mm}$$

This result could have been obtained more directly by using a formula:

Length of one 'millimetre' on the inclined scale $= \dfrac{1}{(A_i/A_e) + \sin \alpha}$

where A_i and A_e are the cross-sectional areas of the inclined portion and the enlarged portion, respectively.

QUESTION 13, p. 168

(a) This is an 'on–off' control system.

(b) As the temperature rises, the rate at which it rises gradually diminishes. Similarly, as the temperature falls, the rate at which it falls diminishes. So the temperature–time graph would have a curved saw-tooth form, as shown in Fig. S.7.

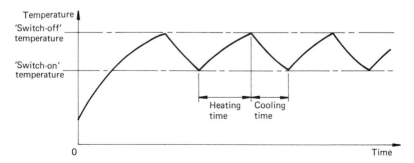

Fig. S.7 Temperature–time graph of the system shown in Fig. 8.2

(c) Referring to Fig. 8.1:

The *control elements* are the switch contacts.

The *reference input* is the force pressing them together, which would be set by the control knob cam.

The *feedback signal* is the reduction in that force, caused by the curling of the bi-metal strip.

Feedback elements is the bi-metal strip.

The *plant* is the electric fire and the air in the room.

The *output* to be controlled is the temperature of the air in the room.

The block diagram is therefore as shown in Fig. S.8.

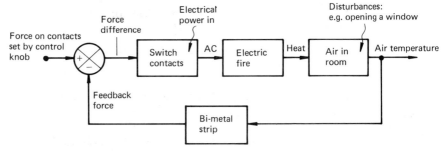

Fig. S.8   Block diagram of the system shown in Fig. 8.2

(d) The corresponding open-loop system would consist of the electric fire, plugged into a mains socket, switched on and left burning.

QUESTION 14, p. 168

(a) Referring to Fig. 8.1:

The *reference input* is the 'full' level FF (dimension y_1).

Feedback elements is the float.

The *control elements* are the arm, lever and inflow valve.

The *plant* is the header tank.

The *output* to be controlled is the actual water level (dimension y_2).

The block diagram is therefore as shown in Fig. S.9.

Fig. S.9   Block diagram of the cistern

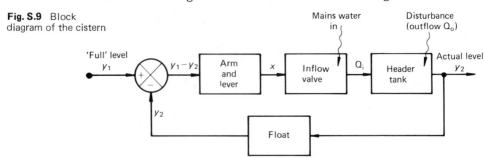

(b) The reference input can only be varied by bending the arm.

(c) The corresponding open-loop system would consist of a tap in place of the inflow valve, set to discharge water at an 'average' rate into the tank. This would cause the tank to overflow at times when there was no outflow, while at other times a large outflow, insufficiently made up by the inflow, would probably empty the tank.

QUESTION 15, p. 170

Referring again to the basic block diagram, Fig. 8.1, let us decide the easy bits first. It seems clear that:

The *output* to be controlled is engine speed.

The *plant* is the engine, including its fuel injection pumps.

The feedback transducer is the rotating masses of the governor.

At this point we have to stop and think. Where do we put the compression spring — in the forward line or in the feedback line? And is the reference input, which we set with the speed control handwheel, a force or a displacement?

To settle these questions, let us consider how we would run the system under open-loop control. We would take out the feedback transducer (the rotating masses). Then, provided we made some arrangement to balance out the weight of the spring and sleeve so that they could not fall down, we would control engine speed directly by continuously adjusting the speed control handwheel. In that case, the spring would just be acting as a distance piece of constant length.

So the *reference input* must be a displacement, not a force. But the output of the feedback transducer is a force, derived from centrifugal force. The spring must therefore be in the feedback line, converting the output of the feedback transducer (force) into a proportional displacement, so that it can be subtracted from the input displacement.

We can now put together the block diagram of the system as shown in Fig. S.10.

 Fig. S.10 Block diagram of all-mechanical speed-control system

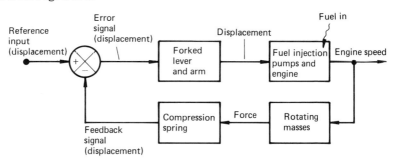

Comparing Fig. S.10 with the speed control system of Fig. 7.3, it is noticeable that the power amplifier of the earlier system is no longer required. This is because the fuel pump control of a diesel engine, or the throttle of a petrol engine require comparatively little force to operate them, while quite large forces are available from the centrifugal force acting on the governor masses. Thus in the all-mechanical system, the power to operate the control elements is provided by the output of the feedback transducer.

QUESTION 16, p. 171

Referring to the basic block diagram of Fig. 8.1:

The *output* to be controlled is the roll position.

The *plant* consists of the main generator, the screwdown motor and the screw jacks.

Feedback elements is the roll position potentiometer.

The *reference input* is the voltage from the reference potentiometer.

Control elements are the power amplifier and the pilot exciter.

The block diagram is therefore as shown in Fig. S.11.

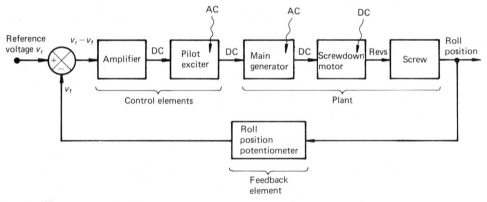

Fig. S.11 Block diagram of thickness-of-plate system

QUESTION 17, p. 207

(a) $$\frac{X_0}{X_1} = \frac{20}{8} = 2.5$$

$$\delta = \ln \frac{X_0}{X_1} = 0.916$$

Damping ratio

$$\zeta = \frac{\delta}{\sqrt{\pi^2 + \delta^2}}$$

$$= \frac{0.916}{\sqrt{\pi^2 + 0.916^2}}$$

$$= 0.280$$

(this agrees with Fig. 10.12 for $X_1/X_0 = 1/2.5 = 0.4$.)

(b)   Length of one cycle on the chart $= 21\,\text{mm}$

$$\text{Corresponding period} = \frac{21}{100}\,\text{mm} \times \frac{\text{s}}{\text{mm}}$$

$$= 0.21\,\text{seconds}$$

∴   Damped natural frequency $= \dfrac{1}{0.21} = 4.76\,\text{Hz}$

'Slow-down' factor due to damping $= \sqrt{1 - 0.280^2} = 0.960$

∴   Undamped natural frequency $= \dfrac{4.76}{0.960} = 4.96\,\text{Hz}$

QUESTION 18, p. 211

(a) (i)   Damped natural frequency, $f_\text{d} = 20/5.1 = 3.9216\,\text{Hz}$

(ii) $X_0 = 20\,\text{mm}$

20 complete cycles are made up of 40 half-cycles.

$X_{40} = 2.5\,\text{mm}$

Using the formulae on pp. 205–6 gives

$$\delta = \frac{1}{2n}\ln\frac{X_0}{X_{2n}}$$

$$= \frac{1}{40}\ln\frac{20}{2.5}$$

$$= 0.0520$$

$$\varsigma = \frac{\delta}{\sqrt{\pi^2 + \delta^2}}$$

$$= \frac{0.0520}{\sqrt{\pi^2 + 0.0520^2}}$$

$$= 0.016\,55$$

(iii) Therefore the undamped natural frequency is

$$f_n = \frac{f_\text{d}}{\sqrt{1 - \varsigma^2}}$$

$$= \frac{3.9216}{\sqrt{1 - 0.016\ 55^2}}$$

$$= 3.9221\ \text{Hz}$$

(iv) The frequency of excitation at which the Q-factor is a maximum is:

$$f = f_n \times \sqrt{1 - 2\varsigma^2}$$

$$= 3.9221 \times \sqrt{1 - 2 \times 0.016\ 55^2}$$

$$= 3.9210\ \text{Hz}$$

When we compare the answers to parts (i), (iii) and (iv) we can see that they are all 3.92 Hz, to three significant figures. The difference between them is negligible because the damping ratio is so small.

(b) (i)   Spring rate $= \dfrac{200}{0.0175}\ \dfrac{\text{N}}{\text{m}}$

$$= 11\ 430\ \frac{\text{N}}{\text{m}}$$

(ii) $f_n = \dfrac{1}{2\pi}\sqrt{\dfrac{\lambda}{m}}$

∴ $3.92 = \dfrac{1}{2\pi}\sqrt{\dfrac{11\ 430}{m}}$

∴ $(2 \times 3.92 \times \pi)^2 = \dfrac{11\ 430}{m}$

∴ $m = \dfrac{11\ 430}{(2 \times 3.92 \times \pi)^2}$

$$= 18.84\ \text{kg}$$

This is the *effective* mass of the beam. The total mass would be about twice this, because only the centre of the beam vibrates with the full amplitude — the ends do not move at all.

Answers to Exercises

CHAPTER 1, pp. 23–7

2 *Transducer*: vaporising liquid; *signal conditioning*: Bourdon tube, link and arm, toothed quadrant and pinion (*amplifier*); *display*: pointer and scale.

3 (a) *Transducer*: Bourdon tube (0.003 33 mm/kPa); *signal conditioning*: link and wiper (5.5), resistance element (0.75 V/mm); *display*: voltmeter.
(b) 0.013 74 V/kPa.

4 (a) *Transducer*: strain-gauge bridge circuit; *signal conditioning*: operational amplifier, power amplifier; *display*: pen recorder.
(b) 180   (c) 0.375 V.

5 (a) *Transducer*: drum (0.573°/mm); *signal conditioning*: gear drive, potentiometer (0.0444 V/°); *display*: voltmeter.
(b) 1:10 gives 258°
(c) 0.00255 V/mm.

7

Range (kN)	0–50	0–100	0–200
Accuracy (% f.s.d.)	0.2	0.3	0.25

Range (kN)	0–500
Accuracy (% f.s.d.)	0.3

Capacity of the machine: 500 kN.

8 b

9 (a) Rotate the zeroing button over the coil pivot.
(b) The pointer must be withdrawn from the pinion shaft and replaced in the correct position.

11 Some of the more common properties for which transducers are available are: acceleration, current, displacement, flow, force, humidity, illumination, magnetic field strength, pressure, rotational speed, strain, temperature, viscosity.

12 (b) ± 1.25 % f.s.d.
(c) (i) ± 2.3 % and (ii) ± 4.8 % of indicated value.
(d) hysteresis

13 (b) $D = 0.643M - 11.4$
(c) ± 1.2 mm

CHAPTER 2, pp. 47–52

3 (a) 0.000 564   (b) 3.67 mm
4 2.11
5 (b) (i) 0.001 117   (ii) 223 MN/m²
(iii) 34.4 kN
7 $A = 0.0603\,\Omega$ (a) 4290 Ω
(b) 12.36 Ω
8 $A = 0.07\,\Omega$, $B = 4900\,\text{K}$
9 (b) C/N (coulombs per newton)
10 14.2 seconds
11 (i) (c)   (ii) (h)
12 (a) (iii)   (b) (v)   (c) (v)
18 (a) (i) (vii)   (b) (iii) (iv)
(c) (ii) (vi)

CHAPTER 3, pp. 85–91

2 (a) 3000 :1   (b) 3 mm
3 1667 :1
4 (a) 6   (b) 34.4 degrees/mm
5 (a) 0.9425 mm
(b) 52.88 (i.e. 53 teeth)   (c) 0.233 %
7 0.001 922
8 130.1 GN/m²
9 (a) 267 MN/m²   (b) 0.001 333
(c) 0.336 Ω   (d) 21 mV
10 One of the many possible answers is shown below. Each bridge circuit must have the gauge numbers shown, but the arrangement of numbers on the square may be rotated or transposed. The essential feature is that the expanding gauges (e.g. 5 and 12) should be on opposite sides of the square, with the contracting gauges (e.g. 11 and 6) between them.

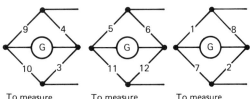

To measure shear force F_s To measure tensile load F_t To measure bending moment M

12 (b) (i) 2.29 V/rad   (ii) 4.80 V
14 (a) 3.96 V   (b) 2.45 V
15 (a) (iv) (b) (v) (c) (iii)
16 (a) Piezoelectric (b) (i) Coulombs
    (ii) Volts
17 (a) 200 s (b) 20 nF (c) 0.1 V
    (d) 50

CHAPTER 4, pp. 104–5

6 (a) (ii)   (b) (iv)   (c) (ii)   (d) (vi)
7 (a) (iv)   (b) (ii)   (c) (vi)

CHAPTER 5, pp. 136–9

3 (a) (i) 7200 and (ii) 4800 flashes per
   minute
   (b) 2400, 1200, 800, 600, 480 and
   400 flashes per minute
4 12 046 rev/min
5 5442 rev/min
8 (c) Zero error
12 2.02% over-reading
14 (c) Charge amplifier

15 (b) (i) Pascals (pressure)
    (ii) Metres (displacement)
17 (a) 2.93 kPa   (b) 104.2 kPa
    (c) 299 mm H_2O
18 (a) 38.8 kPa   (b) 35.9 kPa
    (c) 913 Pa
19 (c) Level the instrument
20 211 mm
21 (b) Cooling water
23 (a) (vi)   (b) (v)   (c) (vii)
    (d) (x)   (e) (ii)   (f) (ix)

CHAPTER 7, pp. 163–5

4 (a) The time lags introduced by the
   components in the loop of a closed-
   loop control system.
   (b) Open-loop gain $\geqslant 1.0$ at the frequency which makes the open-loop
   phase lag 180°.
5 (a) (i) 12.99 Hz   (ii) 0.0019 m/N
   (iii) 79.5°
   (b) (i) 10.53 Hz
   (ii) 156.6×10^{-9} m/Pa   (iii) 189.5°
6 Unstable, because phase lag is 180°
   while gain is 1.178 (i.e. > 1)

CHAPTER 8, pp. 172–6

3

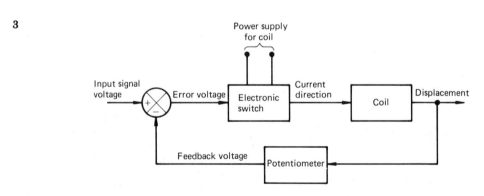

4

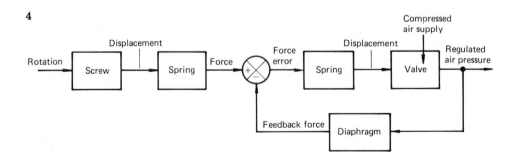

5

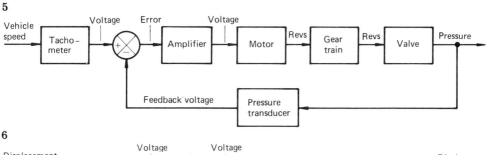

6

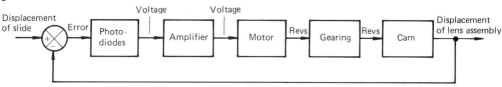

7

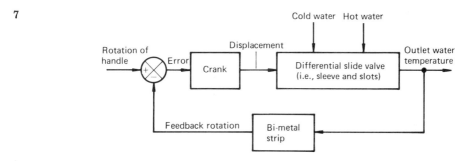

CHAPTER 9, pp. 194–6

2 (c) $0.4s_r - 0.6s_f$
3 See Fig. 9.3, p. 179
4 See Fig. 9.4, p. 180
5 See p. 180
6 See Fig. 9.7, p. 181
8 See Figs. 9.9 and 9.10, pp. 182–3
9 See Figs. 9.11 and 9.12, p. 184
10 See Fig. 9.13, p. 185
11 See Fig. 9.14, p. 186
12 (b) (i) air-to-open   (ii) air-to-close

CHAPTER 10, pp. 222–5

3 (a) 0.330   (b) 0.833 Hz
   (c) 0.883 Hz
4 (c)
5 (a) 0.358   (b) 1.25 Hz
   (c) 1.339 Hz
6 55.7%
7 (a) 0.01103   (b) 3.36 kN/m
8 (a) 0.001600   (b) 4.50 Hz
9 See Figs. 10.15 and 10.16
10 (a) $f < f_n$;   (b) $f = f_n$
    (c) $f > f_n$
12 (a) 0.2   (b) 104 Hz   (c) 0.13
    (d) 170°

13 (a) 0.35   (b) 173 Hz   (c) 0.42
    (d) 150°
14 (a) 1.0   (b) 200 Hz   (c) 0.30
    (d) 113°
15 (a) 2.0   (b) 180 Hz   (c) 0.15
    (d) 104°
16 (b) (i) 2.52 Hz   (ii) 2.08 Hz
    (iii) 45 mm
17 (a) 0.80   (b) 1.67   (c) 0.3
18 (a) 1.48   (b) 103 kg
19 (a) vibrations of very large amplitude.
    (b) (i) increase λ or decrease m
        (ii) decrease λ or increase m
    (c) large amplitude vibrations when
        accelerating through the resonant
        frequency
    (d) damping, or tuned vibration ab-
        sorber

MISCELLANEOUS EXERCISES, pp. 226–8

Answers are not provided, so these
questions may be used for assessment of
students. The answers can of course be
found by reading the book.

Index